Guerrilla Gunsmithing

Quick and Dirty Methods for Fixing Firearms in Desperate Times

Ragnar Benson

Paladin Press • Boulder, Colorado

Also by Ragnar Benson:

Acquiring New ID
Do-It-Yourself Medicine
Eating Cheap
Hard-Core Poaching
Live Off the Land in the City and Country
Mantrapping
Modern Survival Retreat
Modern Weapons Caching
Most Dangerous Game
Ragnar's Action Encyclopedia, Vol. 1
Ragnar's Action Encyclopedia, Vol. 2
Ragnar's Guide to Interviews, Investigations, and Interrogations
Ragnar's Guide to the Underground Economy
Ragnar's Tall Tales
Ragnar's Ten Best Traps
Ragnar's Urban Survival
Starting a New Life in Rural America
Survival Nurse
Survival Poaching
The Survival Retreat
Survivalist's Medicine Chest
Switchblade

Guerrilla Gunsmithing:
Quick and Dirty Methods for Fixing Firearms in Desperate Times
by Ragnar Benson

Copyright © 2000 by Ragnar Benson

ISBN 13: 978-1-58160-119-0

Printed in the United States of America

Published by Paladin Press, a division of
Paladin Enterprises, Inc.,
Gunbarrel Tech Center
7077 Winchester Circle
Boulder, Colorado 80301 USA
+1.303.443.7250

Direct inquiries and/or orders to the above address.

PALADIN, PALADIN PRESS, and the "horse head" design
are trademarks belonging to Paladin Enterprises and
registered in United States Patent and Trademark Office.

All rights reserved. Except for use in a review, no portion of this book may be reproduced, stored in or introduced into a retrieval system, or transmitted in any form without the express written permission of the publisher. The scanning, uploading and distribution of this book by the Internet or any other means without the permission of the publisher is illegal and punishable by law. Please respect the author's rights and do not participate in the any form of electronic piracy of copyrighted material.

Neither the author nor the publisher assumes any responsibility for the use or misuse of information contained in this book.

Visit our website at www.paladin-press.com

Table of Contents

Introduction 1

chapter 1   Barrel Obstructions 7

chapter 2   Sights and Sighting In 21

chapter 3   Lost Parts, Headspace,
                 and Other Bits & Pieces 31

chapter 4   Understanding Gun Metal 49

chapter 5   Ammunition 63

chapter 6   Subcaliber Adapters 83

chapter 7   Fiberglass 93

chapter 8   Guns and Glue 103

chapter 9   Clips and Magazines 111

chapter 10   Repairing Broken Stocks 123

chapter 11   Making Small Parts 135

Conclusion 147

Warning

TECHNICAL DATA PRESENTED HERE, particularly data on the construction, adjustment, alteration, and use of firearms and ammunition, inevitably reflects the author's individual beliefs and experiences with particular firearms, equipment, components, and techniques under specific circumstances that the reader cannot duplicate exactly. It is also the reader's responsibility to research and comply with all pertinent local, state, and federal firearms laws. The information in this book therefore should be used *for academic study only.* Neither the author, publisher, nor distributors of this book assume any responsibility for the use or misuse of information contained in this book.

Preface

My introduction to the gun culture came at a very early age.

LIKE MOST FOLKS IN THE GUN CULTURE, my introduction into this wonderful world of history, steel, and gunpowder was harmless enough.

Wherever Dad got that old Montgomery Ward single-shot, bolt-action hand-cocked .22 rifle with its intensely ungun-like maroon stock is forever lost in history. Perhaps he purchased it for a dollar or two, new out of the catalog, sometime during the dark days of the Depression in 1933 or '34.

On my first outing with the piece in 1949, we shot no more than 10 rounds at wooden fence posts at what I recall to be a range of about 25 feet. Penetration of the bullets into—and through, in many cases—the soft cedar posts was fascinating to me.

I was 10 years old at the time. My brother, who eventually only half-heartedly joined the gun fraternity, was six. But in my case the round was firmly seated and crimped, so to speak. I became addicted to guns and the smell of gunpowder in the morning.

Soon thereafter, I received my first lesson in trajectories and ballistics. "Let's fire a bullet into the moon," said I.

"Too far and too little power," said Dad. "Twenty-two rounds won't likely get past our section road, much less hundreds of thousands of miles out to the moon." (Much later it became obvious that his lectures about shooting the moon exhausted his knowledge of ballistics.)

"What about Uncle Rolf's big noisy Jap rifle he brought home from the war?" I responded. "Surely that gun has enough power to hit the moon!"

"No human has ever made a gun large enough to shoot the moon," Dad patiently explained.

From there it escalated into stuff common to all kids of the era who took a fancy in guns. I scrounged from the dump an old, rusty .22 Stevens with a single-shot falling-block action. It needed a stock, firing pin, and new barrel. Looked like it lived in a mouse nest, the metal was so rough.

I chiseled and whittled a stock of sorts out of a piece of pine 2 x 6. For a firing pin I used a hand drill to auger out what was

Preface

left of the old, rusted, weakened issue firing pin. The replacement was a nail, turned down to size in an electric drill and formed with a common file.

Given our primitive resources, renewing the barrel was impossible, especially for a kid of 10. The gun shot okay but with insufficient accuracy to consistently hit a rabbit or pheasant out to 20 feet. I toyed with the idea of substituting a piece of radio antenna from a junkyard car for the old, rusted barrel. But even at that tender age I realized a loose-fitting smooth bore was a poor substitute for a rifled one, even if it was black and flaky with rust. Additionally I was learning the lesson that the more I shot the barrel, the cleaner it became. To my surprise, accuracy gradually improved rather than deteriorated.

There were a great many more junker guns that we kids brought back to life, including some regular pistols and, of course, our zips and pipes with which we continually experimented. All these childhood experiences turned out to be quite valuable later in life. When I went to Africa in 1967 and '68, there was more than ample opportunity to field every kind of rusted, dented, and destroyed gun imaginable.

At the time, many terrorists lurked on the Kenya-Somali border, so it was imperative that we got some sort of firearms working for personal defense. It was a sorry mess of pottage we came up with, but they kept us alive. And that's what really counts. At the time the Somali *shifta* were mostly armed with World War II Italian Mannlicher Carcanos. Our armaments, cruddy as they were, fairly evenly matched theirs.

During that era we brought an amazing number of junk guns back to life, thanks in no small part to the tremendously skilled and persistent Indians and Chinese who made up the *fundi* class of artisans in east Africa at the time. I saw double rifles re-regulated to shoot to the same point of impact, scopes torn from their mounts replaced, and feats of broken stock repair that absolutely boggled the mind. We removed hopelessly stuck rounds, straightened bent barrels, replaced lost springs,

Many, if not most gun emergencies now occur in the Third World, where gun ownership is tightly controlled. We may be faced with these problems in the United States in the future.

smoothed out dented and destroyed magazines and followers, and made repairs to some pretty hopeless stocks. No doubt more than 30 years later there is still a battered SKS 7.62x39mm with a stock repaired with a piece of hide from a Thompson's gazelle floating around Africa someplace.

Not much later, the process repeated itself in Turkey, and then in the Philippines, Thailand, and Indonesia. We had to have guns for protection; junk was all there was available. Once again, we deployed innovative means and methods as a first step toward securing workable weapons.

In some cases our improvisation actually became pretty sophisticated. Faithful readers will recall accounts of using subcaliber devices jury-rigged into shotguns in the Philippines. We had scrounged up .223, 9mm, and some .38 caliber ammo, but

Preface

we had virtually no rifles or pistols with which to fire it. As a result we tooled out inserts, allowing us to fire these rounds in 12 gauge, single-barrel shotguns. It turned out to be almost terrifyingly effective. As an added plus, the sight of battered old single-shot shotguns raised little alarm among the ruling military. When necessary, we could always ditch the incriminating inserts, even considering the fact that few Filipino GIs really knew what they were.

My early experiences, along with a genuine love for guns and the gun culture and some hard, nitty-gritty experience in tough places around the world, has put me in a position to make a few comments about the skills involved in guerrilla gunsmithing. Seems to me that if I haven't done it personally, I have seen it done by an expert in Somalia, northern Pakistan, or some equally remote location.

Keep in mind that it is only by the most desperate stretch that a real gunsmith would class me as one of his kind. But by reading this book you are going to learn how to patch together and cobble up guns in an emergency situation, not how to be a professional gunsmith.

This book is not about working on guns in situations so primitive that no tools—not even electricity—are available. Even primitive arms builders in northern Pakistan and wartime Vietnam had access to electricity. It's about creatively using whatever simple tools and techniques are available to keep our firearms in working order so we can maintain the ability to defend ourselves and our freedom with them if necessary.

Few countries today permit open, unencumbered ownership of firearms. The United States is quickly becoming like all of the others in the world in this regard. For this and other reasons, knowing emergency gunsmithing procedures is always a very valuable skill for free citizens to have, build upon, and preserve.

Introduction

In times past, every community supported a regular gunsmith. Today there are relatively few in the trade.

Guerrilla Gunsmithing

A CLOSE FRIEND WHO LIVES IN CHICAGO sends guns he wants repaired, rebuilt, or modified 2,000 miles to me for forwarding to local gunsmiths. Unlike in Chicago, four very good, full-time gunsmiths live within 50 miles of me; a fifth and sixth live within 150 miles. The latter two are sufficiently sophisticated that they build and rifle their own barrels from steel bar stock.

Undoubtedly there are still some good gunsmiths in the Chicago area. I say "area" because gunsmithing within the confines of Chicago city limits itself would be a stretch. Extremely restrictive rules and regulations make both the problems of repairing guns *and* dealing with bureaucrats a virtually impossible pain in the rear. Bottom line is that good, competent gunsmiths may live and work in the area, but neither my buddy or I know where they are.

That bureaucratic rules and regulations increasingly discourage a fine, reputable occupation such as gunsmithing should come as no great shock. What bothers me immensely is that so few young men are taking up the trade. (As an aside, I have known of women who repaired chain saws, shod horses, swept chimneys, and repaired lawn mowers, but I have never known a single female gunsmith, although it would seem to be an excellent female occupation.)

Many young guys reading this book probably assume that gunsmiths were always few and far between, but the truth is otherwise. Us older guys can still recall when virtually every small village supported at least one gunsmith. Larger cities might have had seven or eight men who repaired and refurbished guns full time. It was an occupation held in relatively high esteem, and this was in the not-too-distant past.

Today, for a number of reasons, there are fewer and fewer people to whom we can take our guns for repair, maintenance, and refurbishing. Again, part of it has to do with bureaucratic rules and regulations that seem intentionally calculated to drive this occupation underground. (In Africa we did most of our own work because we did not want the authorities to discover that we even *had* guns.

Introduction

When we did contract out our repairs, it was in bits and pieces to rural craftsmen we knew we could trust.) But there are many other forces and trends coming into the picture which, when played out, will make finding a gunsmith increasingly difficult.

First of all, it is generally not as profitable as other occupations. We seem willing to pay gunsmiths only $25 per hour while similarly skilled machinists pull down $40 or more.

Secondly, modern, smokeless powder and noncorrosive ammunition has done away with or mitigated the need for extensive, expensive repairs to many guns. We can fire thousands of rounds before serious cleaning and maintenance is required. When guns do break down, we think first of returning them to the factory for repair. In some cases the gun may be pitched in the garbage like a broken waffle iron. Modern consumption-oriented programming is such that we simply go out and buy new. Ours is a throwaway world, not a repair and reuse one.

Finally, as compared to when I was a kid on the farm, most guns are no longer used as tools. As a result, they do not wear out from constant handling and use. If wear is likely, people generally are financially capable of purchasing several guns to handle the same job rather than relying on one or two patched-up old workhorses.

My point is that not only in a primitive guerrilla situation but also due to the demise of traditional gunsmithing in modern life, we are being forced to handle our own emergency work on our guns and to learn how to use simple means and methods to keep them functioning. Throw in the fact that gunsmithing of some sort may someday become a life or death matter and we begin to understand the significance of the information in this book.

But let's look on the positive side for moment. Presently in the United States several other factors are working in our favor. Many machinists will accept gunsmithing jobs if asked properly. It may not include, for instance, a complex debugging of a Grendel pistol, but it would likely encompass such tasks as drilling and tapping for a scope or installation of a new barrel. Virtually every city has a

Gun owners may be able to find a sympathetic machinist who will handle their work on firearms.

machine and welding shop. Your job is to discover which ones will accept gun work, and which ones will do it competently.

Most machinists won't or can't do wood work. Fortunately, there are dozens of books available that explain in detail how do-it-yourselfers with average skills and patience can handle typical tasks such as inletting and fitting a stock. Additionally, many modern guns are assembled with composite Kevlar fiberglass stocks that are easily replaced when damaged. Composite stocks are also easier to repair with fiberglass and modern epoxies. In times past, much of this work had to be done on a custom, after-market basis by local gunsmiths.

With metal work, modern factory manufacturing techniques have, in large measure, superseded the need for and the capabil-

Introduction

ities of up-country gunsmithing. Factory machines produce sight mount rails, screw holes, sling swivels, and other components so quickly and easily that, for us old guys, it is nothing short of miraculous.

When it comes to the state of modern gunsmithing, there is one more factor that comes into play. I discovered it while working on my recent book on the underground economy. Many good gunsmiths no longer advertise their services or even disclose their presence except to close personal friends. These are not full-time gunsmiths but rather men with day jobs in town. They often work in the evenings only for cash, no receipt offered or accepted.

Like looking for the machine shop that will take gun work, you may be able to find the guy who, for love of the craft, money, or freedom (not necessarily in that order), will do gunsmithing. As I discovered in Africa, Thailand, and elsewhere, these artisans *are* out there. Because of official stigma, they often can't advertise the fact that they can and will work on guns. Realizing that you should root these people out and befriend them now before the hour of desperate need may be one of the more important lessons of this book.

Besides that, the important things to keep in my mind as you read along are these:

1. We are not talking about turning you into a super-skilled gunsmith. What follows are tactics of last resort.
2. You may be better off looking elsewhere for help before engaging in some of these primitive repairs. It all depends on the level of danger you are facing at the time.
3. And this is most important: most places in the world are not friendly toward private gun ownership, let alone gun repair. Our country is rapidly becoming like other countries around the world. In that regard, it is never inappropriate to have this information on hand.

chapter 1

Barrel Obstructions

Guerrilla gunsmiths will be amazed how frequently they will encounter guns with an astounding variety of metal and junk stuck in the barrel.

SLOWLY, ALMOST FURTIVELY, Illian picked his way down the worn trail leading from Mt. Kulau in northern Kenya toward our station. Thick, damp, cottony clouds covered the mountains, adding immeasurably to the misery of the morning.

On the other hand, this was Illian's home territory. He was accustomed to its characteristics and peculiarities. Passing through dense clouds that covered the area 50 percent of the time was no special challenge for him.

Wet, leafy foliage occasionally brushed against the tall, thin man as he almost musically negotiated the wobbly trail. On his shoulder, sometimes balanced on his head, was a parcel securely wrapped in dirty, tattered, native tanned leather. It was big bundle, about 5 feet long and perhaps 10 inches in diameter. Not something a curious bystander would quickly recognize as anything in particular.

This is how Illian and his desperately illegal .303 Enfield Mark III arrived out of the mist at my shop early one morning.

Paramilitary raiders from Somalia were common in the area. First light of dawn was their preferred time. My routine was to be about and ready at that time of day. Not three weeks earlier there had been a raid on Illian's village. Eight people died in an incident so common it did not even make the news. But there was rumor of an illegal gun in the village that was used to help repulse the raid. This was my first chance to actually see it.

Carefully, almost religiously, Illian unwrapped his valuable parcel. Native Kenyans in the region could not legally own a gun under any circumstances, even though they were desperately needed to protect otherwise defenseless villagers against well-armed Somali irregulars. Illian knew full well that I sympathized with his circumstance and would help him in any way possible. It was I, after all, who had once tricked the local district commissioner out of a few rounds of precious .303 ammunition and, on a complete hunch, turned them over to Illian.

His problem that morning was truly a matter of life or death: the .303 Enfield was inoperative. Under normal circumstances we would have taken one look at the battered and

Barrel Obstructions

Nasty old cartridges are frequently wedged tight in gun barrels. When the barrel is full of such junk, the situation becomes very difficult.

abused relic and thrown it in the trash barrel. That antique rifle was so old and worn it didn't even have many usable parts. But under the circumstances, given the incredible dangers faced by Illian and his neighbors, we had to fix it.

The problem with the rifle was immediately obvious. In a desperate attempt to use every precious round of ammunition in his possession, Illian crammed a green, corroded cartridge in his gun. It not only failed to fire but became firmly stuck in the chamber. The bolt would not pull the round.

Believing that any metallic object might spark off the round, he tried valiantly to drive the cartridge from the chamber with a wooden arrow shaft. All he succeeded in doing was driving the weak, wooden shaft firmly down into the barrel, where it broke off even with the muzzle.

An obstruction can sometimes be shot from a gun by pulling a bullet from a cartridge and using the powder alone to dislodge the foreign material.

It was a real mess—but a mess that will be fairly common to guerrilla gunsmiths. Finding weird stuff jammed into barrels of guns is one of the most common situations you will face.

What follows is a rundown of various field-expedient methods to try in a situation like this. Not all of them are applicable to every circumstance, but using some combination of these methods will eventually clear gun barrels and chambers, bringing them back to active service.

BLOW IT OUT

This first technique was popular with American and British doughboys during World War I. It is simplicity itself, but it is only practical under a few narrow circumstances.

Barrel Obstructions

A new bronze brush can be inserted in the chamber containing a separated ring of brass, enabling the guerrilla gunsmith to extract the brass.

In this instance, the barrel is plugged with mud, snow, a stick, or whatever. It may even be a bullet from a defective round, and there is no steel cleaning rod with which to clear the barrel. The following technique should only be attempted with lighter obstructions that are wedged near the muzzle. Do not attempt it if you suspect a double blockage or if the obstruction is lodged tightly right in front of the chamber. Resulting pressures could damage the weapon and, quite possibly, you.

Lay a loaded round of ammunition on a rock, piece of steel, truck bumper, or whatever is at hand. Tap the neck of the cartridge round and round repeatedly with a hammer, rock, or other hard object. This will loosen the cartridge's hold on the bullet enough to enable you to pull it out. Remove the bullet, being careful not to dump any of the powder on the ground. For

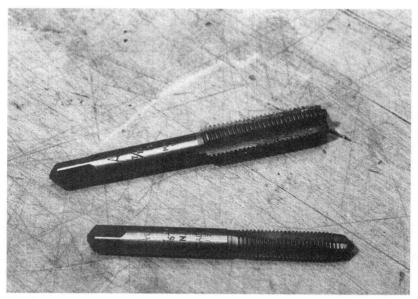

NS Fine taps in 7/16 and 5/16 inch can sometimes be used to remove stuck separated brass.

safety's sake, you could cut the powder charge in half at this point and see if it is sufficient to dislodge the obstruction.

While pointing the gun up in the air to keep the powder in place, carefully insert and fire the bulletless cartridge. This will drive obstructions up to three times the weight of the original projectile safely out of the barrel. There are few if any problems with damage to the gun or user safety with this method, although it is always wise to protect your ears, eyes, and exposed skin whenever firing powder under any circumstances. In fact, if time and circumstances permit, it is advisable to secure the weapon and fire it remotely with a long string, though this may not be possible in a true emergency. Either way, it is a wonderful, field-expedient method to drive obstructions from a barrel that cannot otherwise be cleared with tools on hand at the time.

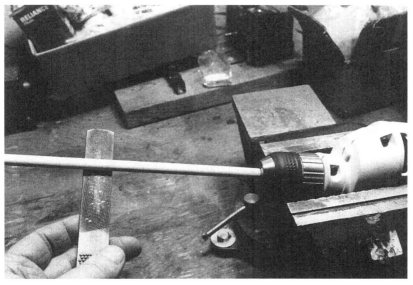

A standard electric drill can be used as a type of lathe to polish down hardwood dowels.

COOK IT OUT

Cooking off an errant round is a field-expedient way to retrieve a stuck cartridge with a faulty primer. It is often used with balky muzzleloaders that refuse to fire. In this example, the gun's action is completely disassembled from its stock and forearm and set against a stump or post in a place outdoors where the discharged round can fly harmlessly into an earthen bank or other solid backstop. A medium flame from a propane torch is then applied to the breach until the round cooks off. About 15 minutes of heating is usually required. This level of heat will not affect the temper of the steel as long as you keep the flame backed off from the piece and are careful not to let the barrel and action get red hot.

Light oil or WD-40 can be used as hydraulic fluid to remove barrel obstructions.

Cooking off a stuck round can be done either with the bolt removed or locked up ready to fire. When done on cartridge firearms with the bolt removed, the cartridge case blows harmlessly back intact out of the breach about 50 percent of the time. If the barrel is obstructed too, cooking off must be done from an open breach. The biggest problem with this method is that when the cartridge case doesn't blow out it will often rupture its base, jamming the brass casing fast in the chamber. If so, try this next method . . .

PULL IT OUT

For whatever reason, spent brass cartridge cylinders with their bases torn off are common obstructions that need to be

Barrel Obstructions

Pour sufficient light oil into the obstructed barrel so that only a small section of hardwood dowel is needed as a hydraulic dam.

cleared. Machine gunners often carry a special tool with which to extract these cases. Revolver shooters are also sometimes faced with them, especially when shooting reloads.

Half the time this separated brass can be removed by inserting a new bronze cleaning brush into the chamber that is one size too big for the job. When reversed, the bristles will dig into the soft brass case, pulling it out. A little lube in the form of Liquid Wrench, brake fluid, or diesel oil may help, but these fluids are not often readily available in the field.

If this doesn't immediately work, place the whole firearm in the freezer for 12 hours. Differential shrinking between the brass and the steel gun will usually allow a release with the brush technique. One time I placed a barreled action with stuck round into a bed of dry ice, which quickly released the brass.

A hardwood dowel about to be pushed into an oil-filled, plugged rifle barrel.

PUNCH IT OUT

When a simple cleaning brush won't pull a ruptured case, try a standard national-fine-thread machinist tap. Making certain the tap is small enough not to engage the chamber barrel metal, gently twist the tool into the cartridge neck. Done carefully, without great force, the tap threads will bite into the brass, allowing it to be punched out from the muzzle end with a regular steel cleaning rod or piece of steel drill rod. A standard screw extractor may also work.

Use a 7/16 tap for most .30 caliber cartridge bodies or a 5/16 tap when working to engage the neck of the brass. If the tap and brass become stuck and cannot be pushed out from the muzzle, it can often be pulled out from the breach with a small

Barrel Obstructions

A common grease zerk can be installed in place of a nipple in muzzle loaders. Grease from a regular automotive gun can then be used to force out obstructions.

vise grips and hammer. Grip the tap with the vise grips while tapping it to the rear with the hammer.

PRESSURE IT OUT

There is one other field-expedient method that might succeed when all else fails. This one is especially valuable when there are other bits and pieces of junk in the barrel.

A steel rod is turned on a lathe or drill press to almost exactly the diameter of the gun bore. If a good grade of hickory rod of the correct length is available, it can be used instead. Like the steel rod, turn it so that it just fits down the barrel.

Half-fill the barrel with light, watery oil, then sharply punch the rod down into the barrel from the muzzle, creating hydraulic

Guerrilla Gunsmithing

pressure against the stuck round. Even if there is other material in the barrel, this pressure will run past it to push against the stuck cartridge.

This is an exceedingly simple procedure if a good, stout rod exactly the size of the barrel can be found or made. I have considered using a heavy piece of felt cloth to seal the hydraulic ram but have never actually tried this.

I once used this technique to remove a stuck round out of a rifle into which a short piece of aluminum cleaning rod had also been wedged. The owner had placed the rod ahead of a piece of steel drill rod to try to remove the round. As a result, the steel riveted the aluminum into the barrel just ahead of the stuck round. I dumped hydraulic oil in the barrel above and around the aluminum rod chunk, which allowed me to place enough pressure on the round to drive it out enough to be pulled from the rear.

A variation of this method works on muzzleloaders. In this instance, something is stuck in the barrel someplace. It is also impossible to remove the rust-frozen breach plug on the gun you are dealing with, or it has a permanently welded plug, so punching out the obstruction end to end is not an option.

In my case, it was a bunch of common carpenters' nails forcibly jammed into the barrel ahead of some very solid patching. Apparently the owner forgot to put a charge in the barrel before ramming in the projectile. The challenge was to remove the obstruction and get the old blunderbuss back in action to defend his family.

Using a propane torch, I brazed a common automotive grease zerk in place of the cap nipple and screwed this assembly into the breach of the gun. It took a whole tube of grease, but eventually hydraulic pressure from the heavy grease gun forced all of the junk out of the barrel. If you try this method, flush the obstruction with water before putting the torch to the piece to insure any powder in there won't go off.

• • • • •

Barrel Obstructions

But how did we proceed on Illian's old warhorse of a rifle? For various reasons, none of the above techniques had much application to his problem.

First I applied as much 10-weight oil and WD-40 as possible around the wooden shaft. I let everything soak for a day while the gun remained hidden in my paint locker. Next I drilled as far as possible into the wood with a quarter-inch bit. Into this hole I turned a 2-inch, #6 wood screw and pulled on it, but I couldn't budge the stuck shaft without danger of damaging the barrel or just removing the 8 inches or so of wooden shaft nearest the muzzle.

The next solution was to use a propane torch to heat the barrel, especially down just ahead of the breach. I got the barrel hot but not red hot, just enough to make the wood smoke pretty good. It was a cautious procedure since I did not want the round in the barrel to cook off, so as an added precaution I wrapped some wet cloth around the breach to keep it from overheating. After that, the slightly charred wooden rod pulled out fairly easily.

Using a piece of good steel 1/4 inch drill rod and working down from the muzzle, I punched the bullet into the case. It still wouldn't budge, but I didn't work it hard at this moment. Instead I filled the bore with 5-weight (sewing machine grade) oil and kerosene.

There was little to no danger of the round going off as a result of my punching on it from the muzzle, but the oil quickly killed the powder and perhaps lubed the base a bit. As it worked out, little of the oil actually penetrated between the breach wall and the green, corroded brass.

After another day of soaking, I again punched down on the cartridge from the muzzle with the quarter-inch steel drill rod. I could have tried to expose the cartridge in the breach and punch through its base from the side in an attempt to pry it out, but we left this as a last option. As it was, the case finally, very reluctantly came out. Neither the breach nor the bore of the rifle were damaged beyond what one would normally expect from a gun of that experience and age. Part of the problem with the stuck cartridge was caused by the pitted, unsanitary chamber, but one has to expect these sorts of things in those places.

I don't know whatever became of Illian and his beat-up old .303 Enfield. For all I know, the rifle is still wrapped in rags and scrap leather, hidden under a floor mat or pile of firewood in his hut in northern Kenya.

chapter 2

Sights and Sighting In

Long warehouse in Mexico where the author sighted in a Ruger Mini-14 with one shot. Residents and workers in the area were not disturbed.

Guerrilla Gunsmithing

NOW AND THEN WHEN LOS MOCHIS, Mexico, native Pedro Alverez is sure nobody important is watching, he will reach into his pickup truck's secret compartment, pull out his Ruger Standard auto pistol, crank down the truck window, and blaze away at ducks and doves sitting thereabouts. Alverez really does have to be careful about his shooting, because all firearms in Mexico are tightly controlled. As a practical matter, private, legal gun ownership south of the border is impossible.

Many, many years have elapsed since we motored around the Los Mochis area together, looking at growing corn, cucumbers, peas, and tomatoes, so his story can probably be safely told now. Pedro is an older guy who learned to shoot and to love guns well before onerous firearms restrictions were put in place in Mexico. The stifling circumstances under which he lived were firmly brought home to me when he decided to sight-in the scope on his new Ruger Mini-14 rifle.

It would have been a virtually impossible assignment given all of the obstacles had I not recalled a one-shot technique we used as kids. It is extremely useful not only when faced with threatening authorities but also because in many survival or Third World situations, the simple fact is that ammunition of any kind is dear. In Pedro's case, he simply could not afford the six or eight rounds needed for a typical sight-in project.

We secured the use of a closed-in warehouse owned by a close friend. The facility was used to store carton boxes used by the tens of thousands in the region to ship winter produce north to the gringos. We considered the fact that noise is something like light—it is best contained when all surrounding cracks and openings are closed, preferably with some fairly heavy, thick material. In the case of old barns in which we frequently fired as kids, not every opening could be closed off, so we had to rely on space. In other words, shots fired far away from a smallish opening in a large barn won't transmit outside for any great distance.

Inside the warehouse we closed all of the doors and windows and piled cardboard boxes against any openings that might

Sights and Sighting In

transmit sound. Our rifle range was nestled down inside piles of cardboard, some up to 30 feet high.

Using bright orange bailer twine, we tied Pedro's Mini-14 securely to the top of a small, wooden piece of furniture. It was kind of a file cabinet made out of heavy 2 x 8-inch lumber that normally sat behind the dispatcher's desk. We stacked 50-pound bags of fertilizer on the bottom shelf to render the whole affair solid as the proverbial rock.

We next placed a paper target out in front of the stabilized rifle exactly 50 yards. On the paper we drew a relatively small dot with a Magic Marker. We then moved the paper rather than the rifle to accomplish our alignment. As the whole thing now sat, the rifle was secured rock solid in one place, aimed exactly at a target dot at a measured 50 yards.

Securing a proper backstop to catch the bullet was as or more difficult than the other parts of the project. It wouldn't do to have an errant bullet go zipping out of the warehouse into the heavily populated areas immediately surrounding it.

After much fooling around we settled on a standard 80-pound bale of hay set on its side so that the rounds had to penetrate the smooth side rather than the cut side. After the round passed through the hay, it would be caught and held by a heavy piece of old carpet hung 15 yards behind the bale. In passing through, the round would lose enough velocity so it would be stopped by the carpet. Had we fired very many rounds, this system would not have succeeded. Successive bullets in a limited area in the bale would have chewed away the hay till it offered very little resistance. It wouldn't have been long until they retained enough energy to penetrate the carpet beyond. Again, not a good situation, given the many people living in the immediate area. (A third backstop placed behind the carpet certainly couldn't hurt.)

The round we fired echoed nicely about the cavernous warehouse but did not seem to bother the night watchman sitting at the front fence. The warehouse was standing about 60 to 70 yards from a camp containing hundreds of sleeping field and factory hands, and as far as we could determine, they either never noticed the noise or a single shot was insufficient to alert them.

The shot hit the paper left of the mark about 2 inches and up high about 3. I checked the scope; it was still perfectly aligned. Recoil from the .223 round is relatively light, so the rifle was not moved off its original point of aim. Rifles with heavy recoil that are jarred off alignment can be physically moved back to the original point of aim and securely fastened down again. In either case, one must be absolutely sure the original point of aim is maintained before fooling with the scope or the sight adjustment in the case of open sights.

Next, we measured up from the point of impact on the paper target 1 1/2 inches and placed a small mark there. Carefully, so as not to physically move the rifle or scope, I cranked it over to the new spot on the target above the actual point of impact. The scope was now sighted in.

This sight adjustment must be done very carefully. Once started, you cannot go back if the rifle is moved accidentally. Also, it will not work with battered old hulks such as Illian's Enfield mentioned in Chapter 1. His rifle probably would not shoot to the same point even over modest 50-yard ranges.

One-shot sight-ins are best done with scoped rifles but can be successfully accomplished with open sights and even on pistols. It is really tough to sandbag handguns properly, but the job can be done.

Knowing the nature of gun nuts like myself, great debate will likely result over my recommendation to sight in 1 1/2 inches high at 50 yards. Please note that this is a general, easy-to-recall-and-implement rule that mostly applies to hunting and military-type situations. It is not sufficiently accurate for bench rest or target shooters.

At plus or minus 1 1/2 inches high at 50 yards, most modern high-power cartridges—including .223, .308, .30-06, and .300 Winchester Magnums—will shoot 3 inches high at 100

Sights and Sighting In

yards, reaching their maximum high trajectory of plus or minus 3 1/2 inches at about 160 yards. Somewhere around 270 yards they will be dead-on again, maintaining their lethal trajectories to about 300 yards. Beyond 400 yards, most bullets will drop 12 or more inches.

This is not a terribly precise measure, but it is sufficient as a general, easily remembered rule of thumb. Past 300 yards most of us are pretty inaccurate shots anyway, even assuming we could correctly estimate these distances.

The warehouse we used in Mexico was sufficiently large that by moving a conveyor and several other pieces of equipment, we were able to shoot 100 yards indoors. A second shot 20 minutes later both failed to stir up the natives and confirmed our sight-in techniques: the shot was about 3 inches high on the target.

MORE ON BACKSTOPS

My favorite backstop is a 1-inch sheet of steel set on a 45-degree angle into which the rounds are fired. In theory, the bullet hits the angled steel plate and ricochets harmlessly down into the ground. Most of the bullet's energy is lost upon first impact. Many of these "traps" are set over a modest 4-inch-deep box of loose sand into which the bullets are caught.

A trap of this type takes enough energy out of rounds that they do not harm cement floors when set up directly over them. All that is usually left of the bullet is some small shards of jacket and core, not even enough mass to pockmark floors.

My personal bullet trap is made from four 20-inch segments of Euclid dump truck spring welded together side by side. Solid .458 and .338 rounds don't even dimple the plate, although they bounce it around a bit. Occasionally the welds will crack.

A swinging steel plate makes another type of effective backstop, so long as the plate is free to swing backward when the bullet hits it. Obviously higher powered rifles and pistols require

A heavy plate of steel set on a 45-degree angle makes a quick, efficient bullet stop.

heavier steel plates. Small .22s will fail to move the plate, causing ricochets, while larger guns may shoot right through.

LASER SIGHTING

Laser Sight Systems manufactures and sells a cartridge-like affair that is really a laser bore sight. The assembly, which looks and acts like a cartridge, is placed in the chamber of a rifle to be sighted in. The little gizmos are offered in .223, .243, .308, .25-06, .270, .30-06, .300 Winchester Magnum, 7mm Rem Mag, and .30-30. When the trigger is pressed, they project a laser beam out to 125 yards, which gives some idea regarding possible point of impact. Theoretically one can do a no-shot sight-in using one of these $100 units.

Sights and Sighting In

A heavy steel swinging plate can be used to stop high-power rounds.

My experience with these devices has been mixed. While bullet impact may be on the paper, in real life individual guns seldom shoot to their actual point of bore sight. It will still take a round or two from the gun to determine where it is actually shooting. If you're still interested, Federal Arms of Fridley, Minnesota, offered these in their catalog last time I looked.

DEALING WITH BENT BARRELS

Especially in a military context, I have frequently run into guns with bent barrels. Some are bent so badly it is obvious they will not zero with available sights. What to do?

For starters, pull the bolt and/or carrier assembly, giving a clear view through the barrel. If necessary, hang a piece of

Rifles or shotguns that shoot off target can sometimes be corrected by straightening or bending the barrel a bit.

Sights and Sighting In

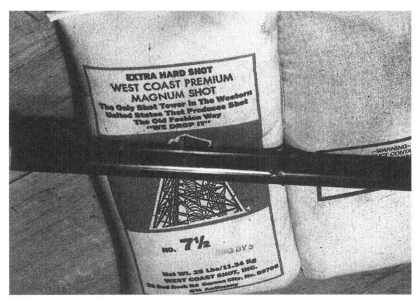

Bent barrels can be straightened by striking them sharply across a bag of shot. Note the kinked area on the barrel.

absolutely straight thread or light wire down on a pane of glass. The shadow cast by this thread or wire down a shiny barrel will look crooked if the barrel is actually bent. (I got especially good at this business when I was a kid installing silencers on our squirrel rifles.)

To begin the unbending process, wrap the muzzle end of the rifle in thick rags and place it in a heavy bench vise. Pull as firmly as possible while tapping with a lead or brass hammer till the barrel moves slightly. It will bend a bit. Until you get into the process you may have to bend and re-bend again till it comes into perfect alignment.

Another excellent, workable method involves laying a 25-pound cloth bag of small lead shot flat on the workbench and vigorously slapping the barrel down on the shot in as near a horizontal position as possible. This method works equally well on shotgun and rifle barrels.

There is, however, an additional problem when straightening shotgun barrels—they want to kink when brought back to true center. Here is how the Indian gunsmiths in primitive East Africa handled this somewhat delicate repair problem.

Take some coarse, heavy rags or preferably a wooden plug of exactly the same diameter as the shotgun barrel. Place the rags or plug in the barrel well below the kink, then fill in about 3 inches on either side of the damaged area with very fine, clean sand. Close it up with a good, stout plug on top of the sand. Both plugs must be very secure, suggesting use of hard wood.

Proceed to straighten, either by striking on a bag of shot or tapping with a lead or brass hammer. The packed sand seems to keep the barrel from kinking on its return to a straight position. This precaution is especially important if it's a very bad bend or you're dealing with a thin-walled shotgun barrel.

Straightening bent and damaged barrels is a common gunsmithing problem in places where conditions are primitive. Readers who must get "down and dirty" in some Third World locations will often face this task.

chapter 3

Lost Parts, Headspace, and Other Bits & Pieces

Guns used under guerrilla conditions frequently lose parts because they tend to be civilian models not made for such harsh duty.

Guerrilla Gunsmithing

AT LEAST TWICE A YEAR some desperate soul shows up on my doorstep with a box full of gun parts, the result of his taking a gun apart that he cannot reassemble.

Reminds me a bit of the years I spent clerking in a big city gun shop. Whenever a World War II widow brought in a Mauser or Luger war trophy, we disassembled the pistol before making an offer to buy. Looking at the great jumble of parts on the counter definitely kept price inflation in check. We kidded about this technique extensively, but the truth is that we always reassembled the gun for the lady. She had enough grief to deal with without our adding to it by depreciating part of her estate.

No doubt the grandest reassembly project I was ever involved with occurred in the early days of Vietnam. Jack Baker's brother had sent him a big box topped up with every part for an M1 Garand, Thompson Model M1A1 submachine gun, M2 carbine, Schmeisser submachine gun, M14, M16, Colt 1911 .45 pistol, AK-47, SKS, and even a PPsh 41! Absolutely every part was disassembled. Barrels were all unscrewed from their receivers, and even sight assemblies were taken down. Putting them all back together was definitely a labor of love, taking most of one winter to complete.

Getting all of the tiny springs and detent balls correctly matched in that great jumble of parts was a major challenge. That—and the fact that we managed to lose some of the real small stuff without really knowing what parts were gone, if they were even there in the first place—contributed to our frustration. Eventually we persuaded Jack's brother to send the little stuff stuck to pieces of tape.

Exploded firearms drawings were uncommon in those days. The best we could do for a reference was W.H.B. Smith's tome *Small Arms of the World*. Now edited by Smith's son, Joseph E. Smith, *Small Arms* is still available and is still invaluable for understanding military small arms, though it doesn't always show absolutely every tiny part for every gun. The problem is that guerrilla gunsmiths rarely will have a copy of the book out

Lost Parts, Headspace, and Other Bits & Pieces

A box of mixed gun parts can be a real puzzle without an exploded diagram or set of instructions for reference.

in the boondocks when they desperately need it. The only alternative is to study and practice on military weapons that you may encounter now so that you become familiar with basic concepts, which will often prove invaluable later.

Fortunately, hundreds if not thousands of exploded firearms drawings are available today. *Gun Digest* has an entire book of almost 500 exploded drawings that they add to and republish every few years. The good folks at the National Rifle Association publish exploded firearm diagrams virtually as fast as new gun designs come on the scene.

Like Chinese ring puzzles, every gun has a "system" to take it apart and put it back together. Some guns, such as Winchester Model 24s or Model 40 autoloader shotguns, are extremely difficult to reassemble even if you do know the system. Unlike mil-

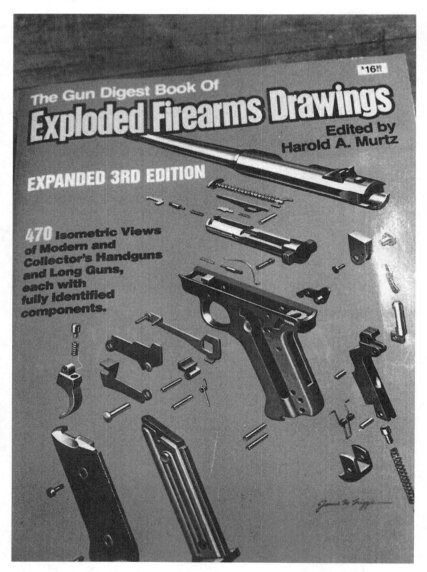

References on firearm parts and design will be invaluable for guerrilla gunsmiths.

Lost Parts, Headspace, and Other Bits & Pieces

itary guns that are specifically made to be field-stripped and cleaned by anyone, these guns are assembled at the factory using a special jig that holds various pins and springs so that they can be slipped into place without damage.

In the case of Model 40 shotguns, Winchester used to recommend that they be returned to the factory for repairs. Those who did so found that the company didn't even try to make repairs; they simply kept the gun and exchanged it for a new Model 12 pump. (Explains why we seldom see Winchester Model 40 autoloaders in collectors' hands today.)

As a practical matter, we are going to have to field whatever guns are on hand, even if it's a cranky old Model 40. If it is something like a Model 40, we can only hope that we have exploded drawings on hand, patience is one of our virtues, we have had time to acquire prior experience, and the situation is not life or death. Like the man said about pipe guns, perhaps the most we can expect is the ability to use them to get some real guns!

Almost all guns designed and produced after 1911 use coil rather than leaf springs. Also, guns made in the last 80 to 90 years are usually characterized by ease of field maintenance. The assembly and disassembly system for many of these guns does not require force or even very much dexterity.

Given their proliferation during the many bush wars that were part of the Cold War, there is a good chance the guerrilla gunsmith will encounter relatively simple military firearms to work on. AK-47s, I understand, hold the world record for number produced. They must also hold some sort of simple design record for the fewest number of parts required to make them operational. There is also the matter of the record speed at which AKs deteriorate in hard service, but no matter—there are millions around, and they are the benchmark for simplicity. Even AR-15s, once you get on to their basic operation, are relatively easy to deal with.

Modern military weapons tend to be simple to work on, but a few designs such as the SPAS-12 are virtually too complex to work on.

THE SMALL STUFF

The biggest problem always involves the smallest parts. These are the small beads and detent balls and springs that fly here and there. These parts, as Jack Baker discovered, are easily lost, but they are generally not tough to replace from off-the-shelf stock.

I have learned to work over a thick terry cloth towel when dealing with small parts. At those times when springs may dislodge, I put a cloth on top of my work to catch such flying parts. Either that or I work in an open, clean room with a smooth cement or linoleum floor.

Through the years my biggest problem with guns has been lost or damaged springs. Modern springs seldom wear out, but they may hopelessly kink in the slide during reassembly or will-

Lost Parts, Headspace, and Other Bits & Pieces

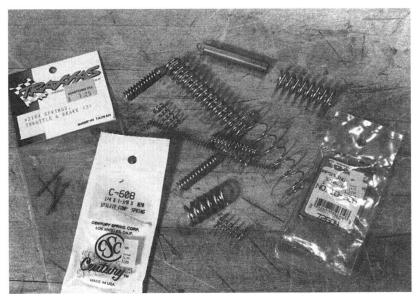

Most springs used in guns today are stock, off-the-shelf coil springs.

fully slip out of their detent whereupon they fly out the third floor window down onto a passing beer truck. But fortunately, there is help.

Coil springs used in firearms these days are similar to stock springs taken off the shelf from existing supplies. Springs are cataloged by diameter, length, and in some instances size of the wire from which they are constructed. A page from an old machinist supply house catalog from what I believe is now a defunct outfit is shown for information purposes only. (Don't try to order from these people. It is only an illustration of how springs are identified, ordered, and priced.)

Locating a supply of springs for possible use in firearms can be a hassle. Grainger Supply carries a 156-piece assortment box in its latest catalog. List price is only about $30, but my guess,

Springs

Music Wire & Stainless Steel Compression Springs

Printed catalog prices are for quantities of less than 25 each per size. For quantities greater than 25, please contact our sales dept. for current prices.

Compression Spring Kits. Contain 252 compression springs in 126 different sizes (2 of each spring listed in table below). Packed in compartmented box; sizes marked in cover.
Music Wire Kit. Cadmium plated. Meets MIL spec. QQW 470.
No. 9434K10 .... NET EACH $83.20
Type 302 Stainless Steel Kit. Passivated finish. Meets MIL spec. QQW 423.
No. 9435K10 .... NET EACH $110.40

O.D. In.	Wire Dia. In.	Free Lgth. In.	MUSIC WIRE No.	NET EACH	STAINLESS No.	NET EACH	O.D. In.	Wire Dia. In.	Free Lgth. In.	MUSIC WIRE No.	NET EACH	STAINLESS No.	NET EACH
.120	.016	1/4	9434K12	$3.36	9435K11	$3.36	.360	.038	7/16	9434K75	$3.36	9435K77	$3.36
		1/2	9434K13	3.36	9435K12	3.36			13/16	9434K76	3.36	9435K78	3.36
		3/4	9434K14	3.36	9435K13	3.36			1 1/2	9434K77	3.36	9435K79	3.36
.120	.018	1/4	9434K15	3.36	9435K14	3.36	.360	.042	7/16	9434K78	3.36	9435K80	3.36
		1/2	9434K16	3.36	9435K15	3.36			13/16	9434K79	3.36	9435K81	3.36
		3/4	9434K17	3.36	9435K16	3.36			1 1/2	9434K80	3.36	9435K82	3.36
.120	.022	1/4	9434K18	3.36	9435K17	3.36	.360	.045	7/16	9434K81	3.36	9435K83	3.36
		1/2	9434K19	3.36	9435K18	3.36			13/16	9434K82	3.36	9435K84	3.36
		13/16	9434K20	3.36	9435K19	3.36			1 1/2	9434K83	3.36	9435K85	3.36
.180	.016	1/4	9434K21	3.36	9435K20	3.36	.420	.038	1/2	9434K84	3.36	9435K86	3.36
		1/2	9434K22	3.36	9435K21	3.36			7/8	9434K85	3.36	9435K87	3.36
		3/4	9434K23	3.36	9435K22	3.36			1 1/2	9434K86	3.38	9435K88	3.36
.180	.018	1/4	9434K24	3.36	9435K23	3.36	.420	.042	1/2	9434K87	3.38	9435K89	3.36
		1/2	9434K25	3.36	9435K24	3.36			7/8	9434K88	3.38	9435K90	3.36
		3/4	9434K26	3.36	9435K25	3.36			1 1/2	9434K89	3.54	9435K91	3.54
.180	.022	1/4	9434K27	3.38	9435K26	3.36	.420	.047	1/2	9434K90	3.38	9435K92	3.36
		1/2	9434K28	3.36	9435K27	3.36			7/8	9434K91	3.36	9435K93	3.36
		13/16	9434K29	3.36	9435K28	3.36			1 1/2	9434K92	3.54	9435K94	3.54
.180	.026	1/4	9434K30	3.36	9435K29	3.36	.420	.055	1/2	9434K93	3.36	9435K95	3.36
		9/16	9434K31	3.36	9435K30	3.36			7/8	9434K94	3.36	9435K96	3.36
		7/8	9434K32	3.36	9435K31	3.36			1 1/2	9434K95	3.54	9435K97	3.54
.180	.032	5/16	9434K33	3.36	9435K32	3.36	.480	.038	1/2	9434K96	3.36	9435K98	3.36
		5/8	9434K34	3.36	9435K33	3.36			7/8	9434K97	3.36	9435K99	3.36
		1	9434K35	3.36	9435K34	3.36			1 1/2	9434K98	3.54	9435K111	3.54
.240	.022	3/8	9434K36	3.36	9435K35	3.36	.480	.042	1/2	9434K99	3.36	9435K112	3.36
		9/16	9434K37	3.36	9435K36	3.36			7/8	9434K111	3.36	9435K113	3.36
		13/16	9434K38	3.54	9435K37	3.54			1 1/2	9434K112	3.54	9435K114	3.54
.240	.026	3/8	9434K39	3.36	9435K38	3.36	.480	.045	1/2	9434K113	3.36	9435K115	3.36
		1/2	9434K40	3.36	9435K39	3.36			7/8	9434K114	3.36	9435K116	3.36
		7/8	9434K41	3.36	9435K40	3.36			1 1/2	9434K115	3.54	9435K117	3.54
.240	.032	5/16	9434K42	3.36	9435K41	3.36	.480	.055	1/2	9434K116	3.38	9435K118	3.54
		5/8	9434K43	3.36	9435K42	3.36			7/8	9434K117	3.36	9435K119	3.54
		1	9434K44	3.36	9435K43	3.36			1 1/2	9434K118	3.54	9435K152	3.65
.240	.038	3/8	9434K45	3.36	9435K44	3.36	.480	.063	1/2	9434K119	3.36	9435K121	3.54
		3/4	9434K46	3.36	9435K45	3.36			7/8	9434K152	3.36	9435K122	3.54
		1 1/2	9434K47	3.36	9435K46	3.36			1 1/2	9434K121	3.54	9435K123	3.66
.240	.042	3/8	9434K48	3.36	9435K47	3.36	.600	.045	5/8	9434K122	3.36	9435K124	3.54
		13/16	9434K49	3.36	9435K48	3.36			7/8	9434K123	3.38	9435K125	3.54
		1 1/2	9434K50	3.36	9435K49	3.36			1 1/2	9434K124	3.54	9435K126	3.66
.300	.022	1/2	9434K51	3.36	9435K50	3.36	.600	.055	5/8	9434K125	3.36	9435K127	3.54
		1 1/16	9434K52	3.36	9435K51	3.36			7/8	9434K126	3.36	9435K128	3.54
		7/8	9434K53	3.36	9435K52	3.36			1 1/2	9434K127	3.54	9435K129	3.66
.300	.026	7/16	9434K54	3.36	9435K56	3.36	.600	.063	5/8	9434K128	3.36	9435K153	3.54
		11/16	9434K55	3.36	9435K57	3.36			7/8	9434K129	3.36	9435K131	3.54
		1	9434K56	3.36	9435K58	3.36			1 1/2	9434K153	3.54	9435K132	3.87
.300	.032	7/16	9434K57	3.36	9435K59	3.36	.600	.067	5/8	9434K131	3.36	9435K133	3.66
		11/16	9434K58	3.36	9435K60	3.36			1	9434K132	3.54	9435K134	3.87
		1	9434K59	3.36	9435K61	3.36			1 3/4	9434K133	3.54	9435K135	3.87
.300	.038	3/8	9434K60	3.36	9435K62	3.36	.600	.072	3/4	9434K134	3.54	9435K145	3.87
		13/16	9434K61	3.36	9435K63	3.36			1 1/4	9434K135	3.54	9435K137	3.87
		1 1/2	9434K62	3.36	9435K64	3.36			2	9434K136	3.66	9435K138	4.00
.300	.042	3/8	9434K63	3.36	9435K65	3.36	.720	.055	3/4	9434K137	3.66	9435K139	4.00
		13/16	9434K64	3.36	9435K66	3.36			1 1/4	9434K138	3.66	9435K154	4.00
		1 1/2	9434K65	3.36	9435K67	3.36			2	9434K139	3.87	9435K141	4.29
.300	.045	3/8	9434K66	3.36	9435K68	3.36	.720	.063	3/4	9434K154	3.66	9435K142	4.00
		13/16	9434K67	3.36	9435K69	3.36			1 1/4	9434K141	3.87	9435K143	4.29
		1 1/2	9434K68	3.36	9435K70	3.36			2	9434K142	3.87	9435K144	4.29
.360	.026	1/2	9434K69	3.36	9435K71	3.36	.720	.067	3/4	9434K143	3.66	9435K145	4.00
		3/4	9434K70	3.36	9435K72	3.36			1 1/4	9434K144	3.87	9435K146	4.29
		1 1/8	9434K71	3.36	9435K73	3.36			2	9434K145	3.87	9435K147	4.29
.360	.032	1/2	9434K72	3.36	9435K74	3.36	.720	.072	7/8	9434K146	3.66	9435K148	4.29
		13/16	9434K73	3.36	9435K75	3.36			1 1/4	9434K147	3.87	9435K149	4.29
		1 1/2	9434K74	3.36	9435K76	3.36			2	9434K148	3.87	9435K151	4.29

Lost Parts, Headspace, and Other Bits & Pieces

looking at the assortment, is that few will match those found in most guns. Grainger has outlets in most large cities in virtually every state if you want to start with them. Also try local machine and welding shops as well as automotive supply stores. If you ask nicely, machine shop proprietors might provide the names and addresses of places from which they order springs. The problem for some guerrilla gunsmiths, of course, is that machine shops are not terribly common, especially in many Third World countries. But look around a bit and hopefully you will be surprised.

Through the years I have purchased three of every kind of spring I have needed to accomplish various gunsmithing tasks. Currently I have hundreds of different ones tucked away in dozens of little parts drawers. Does this mean I can always put my hand on a specific spring when I need it? Not on your life. Fortunately I also live 30 miles from a city with a large machine shop supply store. Seems inefficient to drive 60 miles round trip for a single 3/32 spring that's only 1 inch long, but I have done so on several occasions.

HEADSPACE HEADACHES

Incorrect headspace in one form or another will be a common malady among the guns on which most guerrilla gunsmiths will be working. This is because the guns will be so old and worn that they just won't hold a cartridge properly. No doubt most should be returned to a real gunsmith for major reconditioning, but that may not be an option.

It may be best to consign old reloaded ammo to that specific gun and/or use it only as an emergency backup. Another school of thought says to use your best, freshest ammo when you have excessive headspace, because it will more likely stretch without splitting in the oversize chamber. If there is a rupture, it is a great tragedy because the user often desperately needs the gun to function properly, and tools to clear the blockage may be 25 miles away.

Signs of excess headspace: cartridge on right is split; at left the brass shows a bright ring, indicating stretching.

Barrels can be replaced in most guns on a field-expedient basis if spares are available. More on this later, but first let's take a quick and dirty look at how headspace problems will likely occur. Again, real gunsmiths will recoil in horror at my analysis. But out in the field you will have few options, nor will you likely have the tools and spare parts with which to do much about it.

Headspace has to do with the bolt of the firearm supporting the cartridge in the chamber. After many years of firing, some actions will stretch a bit; conversely, the locking recesses in the receiver that hold the locking lugs of the bolt may gradually flow to the rear, compromising the necessary support the bolt provides to a round in the chamber (i.e., the bolt will not push the cartridge far enough forward in the barrel).

Lost Parts, Headspace, and Other Bits & Pieces

Brass cartridges stretch to fill the chamber when fired; that's what makes them effective gas seals. The problem sometimes is that they don't stretch enough to fill the space, allowing the base of the cartridge to stretch away from the body. Older, brittle brass, cartridges reloaded too many times, and corroded rounds put through worn rifles frequently fail at critical times in this manner. When they do, it's back to Chapter 1 to remove stuck brass from the chamber. Guns *can* be blown apart when gas escapes to the rear, but this is very unusual even with cheap guns made of soft metal.

Real gunsmiths have gauges they use to measure headspace. Guerrilla gunsmiths can quickly determine how badly their guns are out of headspace by noting the extent to which their fired cartridges have stretched when compared to a factory-new cartridge. I always keep one or two factory loads around for comparison. Overly stretched cartridges, for example, will have telltale marks that show up as a shiny ring around the base of the brass. After you have seen examples, you will know exactly what I am talking about.

BARREL REMOVAL

Occasionally, headspace and other problems can be traced back to barrels that have worked themselves loose in their receivers. Usually it happens that barrels are frozen into receivers by years and years of rust and corrosion, but they do become loose sometimes too (often because they were not installed correctly to begin with).

For this reason (and because guns that have had their barrels disassembled from their receivers are easier to camouflage, transport, and hide), guerrilla gunsmiths must know how to take a barrel from a receiver. It may also be necessary to replace a worn, jammed, or burst barrel. When tightening an existing barrel to the receiver, the first rule is always to adjust the barrel, not the receiver. Try peening down the threads on the bar-

Guerrilla Gunsmithing

Guerrilla gunsmiths will have to learn how to take barrels from actions so they can work on these guns and perhaps better hide them if necessary.

rel just a bit. If that doesn't produce a tighter fit, make a washer out of a thin strand of copper wire and affix it on the barrel to take up a few thousandths inch of slack. Use as little wire as possible to fix the problem.

When taking a barrel from a receiver, start by completely disassembling the gun down to its basic barrel and receiver. In a few cases open front and rear sights can be left in place, but if the sight will interfere with securing the barrel in a vise, it should be removed. Scope sights must all be removed before disassembling barrels.

The receiver should always be unscrewed from the barrel, not the other way around. This means that the barrel will be securely clamped solidly in a vise so that the receiver can be turned from the barrel. Barrel clamping could be done in a

Lost Parts, Headspace, and Other Bits & Pieces

A good temporary barrel vise can be assembled with blocks of soft wood set in a large steel vise.

plumber's vise, but scarring would be horrible. Granted this isn't refined gunsmithing, but you should still try to keep dragon bites on the gun to a minimum.

I clamp barrels in a regular bench-mounted, 12-inch, machinist's cast iron vise. Four 1-inch soft pine boards are used as jaw pads, two on either side of the barrel. Cinch the vise down hard enough so that the barrel presses deep into the board about halfway round. This amount of pressure usually keeps it from slipping while you turn the receiver.

Actions should not be torqued so much that they warp or twist while dismounting barrels. If a great number of similar barreled actions are to be worked on, it may pay to make a special turning tool. Otherwise use either a large crescent wrench or a heavily padded pipe wrench. I often cut out a block of

Marks and barrel and action show correct alignment of this screwed-in barrel.

wood or steel that fits nicely inside the action to keep it from deforming when twisted. This is especially important when working on modern military weapons that have lightweight aluminum alloy receiver frames.

In some cases, guerrilla gunsmiths will have to heat receivers to free them up. Use an acetylene torch to bring the metal up to real hot, but not red or even gray hot. But do this only as a last resort after trying Coca Cola, WD-40 penetrating oil, or brake fluid to cut rust and corrosion.

Replacement barrels of the same cartridge are turned back into their receivers until the index marks on the bottom of the action and barrel match. This way, sights will be upright and cartridge feed ramps will be in their proper position. All barrels and receivers that have ever been paired have these index marks,

Lost Parts, Headspace, and Other Bits & Pieces

Modern welders such as this gas-shielded, wire-fed, electronic model, can be of great service to the guerrilla gunsmith.

although keep in mind that a replacement barrel's marks may not exactly match up with the receiver's mark. The headspace will need to be checked and corrected if necessary.

WELDING BASICS

Guerrilla gunsmiths will probably have to own or have access to an acetylene torch. These tools are so universally handy they are common even in the Third World in spite of sometimes horrible logistics problems. For gunsmith work, an acetylene torch is preferred over arc welding, either the old-fashioned stick welders or the modern electronic systems that use inert gas. Versatile torches can be used to weld, cut, heat-expand frozen metal parts, and anneal and harden gun parts.

There are three basic types of welding: forge, electric, and gas. Each method fuses metal pieces by heating them until the melting point is reached and the separate parts flow into each other.

Welding is dramatically different from soldering, silver soldering, and brazing. These techniques are really a type of hot gluing (or surface alloying, to be more precise) using metallic substances to hold parts together. The higher melting temperature of solder, silver solder, and braze means they are tougher to handle, but the finished product will produce a stronger, more durable union than epoxies.

To a large extent, how well the parts bond is dependent on how well the two surfaces have been thoroughly cleaned and then treated with welder's flux. Various fluxes are available for various surfaces. Explain the matter to the man in the machine or welding supply shop. He will make recommendations or look in the Brownell catalog, which lists dozens of special-purpose fluxes.

Most gunsmith work is done with a common, old-fashioned oxyacetylene torch, commonly known as an acetylene torch. Oxyacetylene is a mixture of carbide gas and oxygen. In areas where liquid petroleum (LP) gas is plentiful, such as Mexico, you will often find it substituted for carbide gas. The results are acceptable for many applications, but it's not quite as good, as LP gas and oxygen does not get quite as hot as the real thing. Cutting large pieces of steel this way can be more tedious, and welding generally must be done with the next larger sized flame tip.

Besides tanks, regulators, a torch, and hose, a few other pieces of equipment are required. Goggles, igniters for lighting the torch, and a tip cleaner should be on hand. It also helps immeasurably to have a wheeled rack on which to move the tanks around. In place of tanks, I have seen acetylene torches in Third World countries that were supplied by inner tubes pumped full of carbide gas. The users skillfully squashed the inner tube in order to get the desired gas pressures. Accidents occurred occasionally, but in general these were handy, workable arrangements.

Lost Parts, Headspace, and Other Bits & Pieces

Braze is little more than hot metallic glue, but it can hold high-impact pieces such as this recoil lug on a magnum-caliber rifle barrel.

The greatest challenge to using an acetylene torch is setting the flame. Adjusted too cold and the weld will be pockmarked with gas holes and will be layered rather than fused (i.e., they will not be melted together). Set up with too much oxygen or too much pressure on the carbide side and you will burn the metal away, perhaps destroying the part you wish to rebuild, join, or weld.

There is no alternative to several hours of practice with the welder before attempting to turn the bolt of a Mauser rifle or repair a bent Enfield magazine. Again, torch adjustment is critical. Flame from the acetylene should burn lazily, just at the tip of the nozzle without visible pressure. Turn the oxygen valve up slowly until the yellow color disappears and a neutral blue flame results.

I always attempt to fuse the metal of two parts before adding in metal. Gunsmith work is usually done with 3.5 percent nickel steel welding rod, giving the weld some additional strength and the ability to take a polish. Welds can also be made using black (not galvanized) wire or bits and pieces of similar grade steel.

Brazing is the hottest solder technique and approaches welding for actual field applications. Since the various parts are not melded together, it is still a solder technique. Braze melts at relatively lower temperature and is ideally suited to repairing the sides of magazines or placing choke devices on thin shotgun barrels, for instance.

Having come from the era when we took zinc acid C-cell battery cores and threw them in hydrochloric acid to make our own brazing flux and used number nine fence wire to weld with, modern fluxing solutions, electric welding rods, welding wire stock, brazing compounds, and the like are nothing short of amazing to me. Many of these materials won't be available in an emergency situation, but if you care to practice now with these modern miracle products, look to Brownells in Montezuma, Iowa, for everything you will need.

There are going to be numerous times when a welding torch will be necessary to forge or alter parts, reweld broken components, and heat something so it can be disassembled. Sounds like a big order, but no matter where I was I always found a torch and some welding materials with which to do the work. Most villages of over 5,000 people are home to someone with a welder who can turn out to be surprisingly skilled given the limited tools and supplies with which he must work. I look for these people. When I can't find them, I do the work myself. But not before practicing on similar but unimportant pieces of steel.

Not only for guerrilla gunsmith work but also for a better life in general, I recommend that everyone learn the basics of gas and electric welding. In my opinion, welding skills are as important as knowing how to drive a tractor, shoot a pistol, or smoke meat. It is a basic survival skill that everyone should have. Short courses are commonly offered by rural electric cooperatives, trade schools, and some community colleges.

chapter 4

Understanding Gun Metal

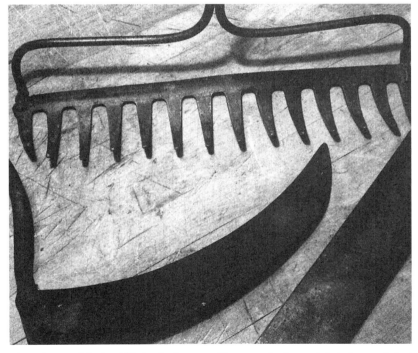

If you know a bit about likely steel composition, seemingly unrelated pieces of steel can be used to construct gun parts.

GUERRILLA GUNSMITHS WILL HAVE TO UNDERSTAND a few basics about composition, chemistry, and other characteristics of various types of steel, if for no other reason than to understand if that old lawn mower blade or ax head, for instance, can properly be used as raw material for a new pistol hammer assembly, firing pin, or stock retainer screw. Here are some guidelines.

COMPOSITION

Steel with a small percentage of carbon is used in most modern firearms as well as in the tools and machines we find around us everyday. All gunsmiths are principally concerned with gun metal comprised of nickel/carbon steels of relatively low carbon content and 3 to 5 percent nickel. These types of steel are not particularly hard, but they are very tough, having great tensile strength and resilience. Low-carbon steel makes the best internal working parts in guns because it can easily be hardened against wear.

The steel used in guns is usually between .84 to 1.0 percent carbon. Barrels, receivers, slides, and bolts are frequently made up of this grade of relatively easily machined material. Common machinists' drill rod, for quick comparison, is about 1.25 percent carbon content. If you're not careful, you may initially believe that any old scrap on hand such as used hacksaw blades or worn engine valves is ideal material from which to construct gun parts, when old pieces of a hoe, for instance, might be much better. This is because of the perception that only "good" steel is used to construct a blade to cut steel or run an engine which, in turn, we erroneously believe would make good gun parts.

The reality is otherwise. Hacksaw blades, for example, with their 1.60 percent carbon content, are very difficult to next-to-impossible to work with. Any hardening will quickly burn out of the part in the process of working, annealing, or hardening. Another practical example: old-fashioned, flat-type gun springs as found in some revolver and double-barreled shotguns are usually

Understanding Gun Metal

These old files and saw blade could be suitable for gun parts if you knew their probable composition.

made from steel with 1.1 to 1.2 percent carbon. In an emergency, this grade steel can be approximated from a piece of metal file, from which a spring could reasonably be fashioned. The guerrilla gunsmith needs to know this and similar real-world equivalents.

Do not confuse these various grades of steel with modern and increasingly common tungsten alloy air-hardened steel. These steels, which automatically harden in open air as they cool, are difficult to work with. Tungsten alloy steel is sometimes found in knife blades, automotive gears, machine tools (including drill bits), and in some newer type guns. (I am told that the gas piston in some modern guns is tungsten carbide.) Common tools, bits, and blades made of tungsten alloy are usually marked as such.

Below is my quick-and-dirty list of common steels and their carbon content. Several pieces of common steel junk are listed

Item	Carbon Content
Drills:	
Wood	.65%
Rod	1.25%
Metal	.80%
Axe	1.20%
Cold Chisel	.85%
Knife Blade	.90%
Lawn Mower Blade	1.00%
Steel Punch	.80%
Screw Driver	.65%
Common File	1.25 to 1.30%
Hoe Blade	.85%
Rake Tines	1.15%
Saw Blades:	
Circular	.85%
Hack	1.60%
Band	.75%
Hand	.85% to 1.00%
Wrench	.85%
Taps	1.20%
Dies	.90%
Flat Spring	1.10% to 1.20%
Junkyard Scrap Angle Iron	.85% to 1.00%
Steel Pry Bar	.85%

Understanding Gun Metal

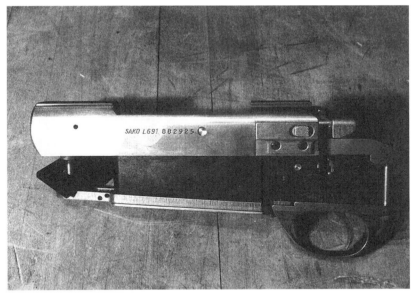

Modern gun metal has 3 to 5 percent nickel and contains little carbon. Gun metal is tough rather than hard.

that obviously don't meet any criteria for use in guns. This is so guerrilla gunsmiths know not to use them, even if it's all that's available. Keep in mind that carbon content alone does not reflect the addition of other materials to the alloys.

Start by recalling that most steel used in guns runs between .84 and 1.0 percent carbon, except internal parts that may be a bit harder and leaf springs that run about 1.20 percent.

To some extent one can determine the makeup of various metal parts by running them over a grinding wheel to look for the kind and amount of sparks produced:

- Nonferrous metal pieces, by definition, don't really spark; generally they are relatively soft, though some nonferrous alloys are harder than high-carbon steel. The grinder will

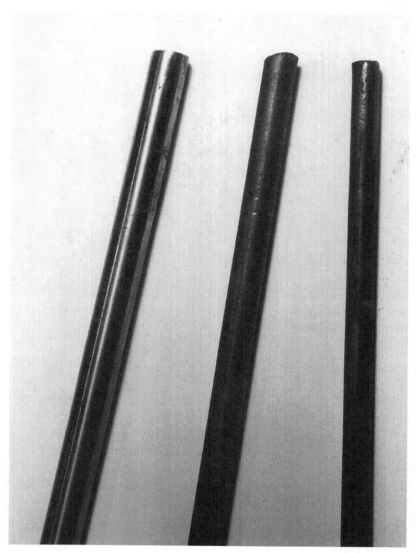

Drill rod on left has a carbon content of about 1.25 percent. It is far tougher than the mild steel rod at center and right. It is also ten times as expensive and harder to find.

Understanding Gun Metal

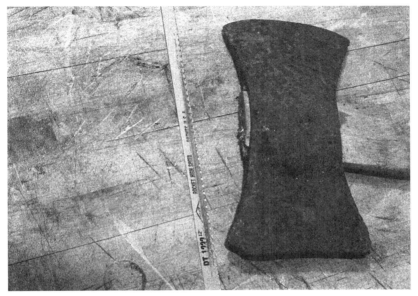

Lower carbon steel in an old ax head may make for better gun parts than a high-carbon hacksaw blade because it is easier to work.

dig in rapidly, producing sparks that are really more shards of flying metal, colored by their native composition.
- Aluminum and lead will produce a silvery white to silvery gray spark-like effect. The shards showering off the wheel may look like sparks, but they are not. There is no fire in them. Also look for the soft metal to color and fill in the grinding wheel.
- If you get a relatively few number of light yellow sparks from 4 to 6 inches from the grinder in long streamers, suspect wrought iron.
- Mild steel is characterized by some—but not many— light yellow sparks with thick streamers close to the grinding wheel. If the metal softens and molds a bit on the wheel, it is a deal giveaway that it is mild steel.

- Many white hot sparks with long streamers indicates that you are grinding on tool steel or gun parts that have probably been hardened.
- Sparks and streamers that are a metallic yellow about 3 inches from the grinding wheel are probably from hardened, higher carbon steel such as some gun springs, wrenches, and drill bits.
- When there is a veritable shower of white hot sparks that seem to pop into being a few inches from the wheel, it is probably manganese or tungsten alloy.
- If the piece slips and chatters on the grinder, it is a dead giveaway that you are grinding on one of the more exotic metals made to resist abrasion and wear.

Other than differentiating between nonferrous metal and high carbon steel, these tests are only an approximation. Old timers who have worked with steel all their lives can quickly tell the difference this way, but hopefully we won't be in the business long enough to get that good.

HEAT-TREATING AND ANNEALING

Sounds strange, but all common steels are heated to both harden and soften them. How this heating is undertaken determines the outcome of the process.

All previously hardened steel must be annealed before it can be successfully worked. Annealing means taking the hardness, or temper, out of the piece. In a guerrilla context, annealing simply entails heating the piece cherry red in a campfire, over an open stove, or with an acetylene torch, then letting it cool very slowly.

Generally it is best if the piece does not come in direct contact with an open flame. Rather than holding the piece over the torch, for instance, it can be heated in a steel box or, more commonly for guerrilla gunsmiths, in an old cast iron skillet over which some sort of steel cap or cover is rigged.

Understanding Gun Metal

Small gun parts are heated to red/yellow hot and then cooled rapidly in water or light oil to harden them.

When heating, be certain the thin portions of the piece do not overheat. Some gunsmiths use a mixture of tin and lead superheated in a pot to bring parts to a uniform temperature. The parts are dipped in the heated lead/tin bath until they are the correct temperature.

As a general rule, there is no problem allowing red-hot pieces of steel on which you intend to work to cool, or anneal, in open air. Fortunate, because this is by far the simplest procedure.

Rural, Third World machinists/gunsmiths have devised a simple, easy method of slow cooling when a part does not anneal by normal air-cooling. Traditionally, such delayed cooling is accomplished by burying the part in powdered lime, powdered charcoal, or even sand warmed on an electric hot plate. Instead, try reheating and then clasping it between two slabs of soft wood. The red-hot part will char its way into the wood, creating an insulating cocoon and allowing it to slow cool while taking the temper out of the steel in the process.

Hardening steel after it is filed and ground into shape is done by reheating and then quickly quenching the part in oil or water. The only absolutely reliable method of uniformly annealing and hardening steel parts is to use a heat-treating furnace. Such furnaces are really tiny ovens capable of reaching very high temperatures. They are used in conjunction with high-temperature thermometers called pyrometers.

It isn't ideal, but for many guerrilla gunsmith applications, it is possible to anneal or harden with either a gas stove or a torch and a small metal enclosure. Results may not be terribly consistent or accurate, but they may have to suffice under the circumstances.

Practice both annealing and hardening with a bit of steel from the surplus scrap pile from which you intend to make a gun part. Most common machine-grade steel must be heated to between 1400 and 1650°F to harden or anneal. At about 500°F the piece takes on a yellow straw color. It will generally turn to a black red at 1000°F. Look for cherry red bordering into yellow again at about 1550°F.

When an open flame is used to heat a piece, the colors will appear brighter. As mentioned, this will be an inexact business without a furnace and thermometer, but there may be no alternative when equipment is limited. Go ahead and take a stab at it.

Judging color is affected by the light under which you are working. In open sunlight the colors are not easily defined. Pitch dark is also not recommended. Evaluate color using a medium intensity, diffused light, such as direct sunlight through a curtain.

An abrupt change in temperature from red/yellow hot to cold is called quenching. Ordinary SAE 10 (not multigrade) motor oil is often used by backwoods gunsmiths to quench small parts. Either salt water or light oil work equally well, depending on what is available. For some strange reason (to me at least),

Understanding Gun Metal

saltwater quenching creates a harder steel part than simply plunging into plain water.

Gunsmiths and especially guerrilla gunsmiths shouldn't fool around with annealing and hardening major gun components such as receivers and barrels. It is a complex and difficult procedure, and results are rarely good. We will only deal with small parts such as hammers, trigger assemblies, sears, sights, firing pins, and perhaps the bolts used to assemble weapons.

WORKING WITH HARDENED STEEL

Probably the best way to learn how to make or alter gun parts and their ancillary requirements to be annealed and/or hardened is to watch someone actually go through the process. With a great deal of patience and perseverance, some very complex gun parts can be made using little more than appropriate scrap, a torch, a file, and a brace and bit. Annealing and hardening to a suitable degree is not difficult for those willing to play with the process a bit.

In an emergency, a simple drill press or even hand-held electric drill can be turned into a kind of lathe/milling machine. Put the piece in the machine and then apply a file. In areas such as Somalia, northern Thailand, and the Khyber Pass region of Pakistan, complex guns are built entirely from scratch using nothing more than a hand-operated breast drill.

Not that many years ago, I had a number of trigger sears made for some older model Browning FN assault rifles at a tiny, rural machine shop near the fabled north Thailand city of Chiang Mai. The proprietor didn't ask what the parts were for other than off-handedly acknowledging that we both knew they were gun parts. His shop was a simple, rough cut, board-sided affair into which he shoehorned a small lathe, drill press, and modest milling machine. Most of the actual work seemed to be done by hand out on the ground in front of the shop while the machinist held the piece between his feet in front of him.

They did a masterful job of constructing the FN parts from broken mechanic's wrenches, which I subsequently learned had a carbon content of about .85 percent. I didn't actually see the heat treatment portion of the project, but I knew they had been hardened by their black color and the hardness of the pieces themselves.

It was a most successful venture, leading me to conclude that before I tried to expend great amounts of time and energy making gun parts with limited tools, I would search around to see if I could find a native craftsman who will expend *his* own skill and energy using limited tools. Some of these guys are truly amazing.

Installing a sight or telescope bases on a receiver may be necessary. Drilling and tapping a hardened barrel or receiver is done by grinding surface hardness from the receiver or barrel so that the steel beneath can be drilled and tapped. Grinding away surface hardening leaves an ugly scar that fortunately is covered by the scope mounting blocks.

Modern alternatives of sorts to drilling and tapping are available. These are covered in Chapter 8 on glues and gluing. Also keep in mind that some gunsmith problems can be overcome with simple screws and strapping, like covering ugly grind marks on Springfield 03s with scope mounts, or the old Indian guns that were bound together with cord. It looks really crummy, but function is not impaired.

USING SURPLUS

Even given unlimited time and motivation, some parts such as barrels and many heavy steel receivers cannot be readily manufactured by hand from scrap steel. It's possible, but doing so requires sophisticated, expensive tools to which we are unlikely to have access. If available, start with ready-made or surplus parts.

Old-fashioned gunsmiths, incredible optimists that they were, always had barrels in their shops bulging with scrap parts and pieces. Afterall, that next gun in for repair might be mend-

Understanding Gun Metal

ed simply with a pin, spring, or screw from that otherwise forgotten wreck in the scrap heap.

Today, parts even within similar models of guns may have changed dramatically over the years. For instance, I have never been able to interchange very many parts from 1873 trapdoor Springfield rifles without a great deal of fitting and filing.

Bottom line? Using existing factory manufactured parts is often the quicker and better solution to most problems. But don't be surprised if fitting is required, and don't be put off by the prospect that you may have to whip out the trusty file and drill to create something completely by hand.

chapter 5

Ammunition

In primitive situations, you may have to use ammunition not made for that specific gun.

WITHOUT AMMUNITION, guns are of little value. Nothing particularly profound about that.

What *is* surprising is the number of different guns that can be made to work with ammo for which they were not originally designed or intended. Likewise, clandestine or emergency reloading of ammo *must* be part of a guerrilla's bag of tricks. There are already plenty of good books on this subject are out there. I will therefore discuss various possibilities and considerations without necessarily going into the nitty gritty of it.

This brief chapter is calculated to get readers thinking about alternatives when one's ammunition supply is otherwise cut off. Most of the suggestions to follow will have gun nuts who know anything about safety hiding under the bench. But if it's an emergency, it is what must be done.

AMMO SUBSTITUTION AND RELOADING TRICKS

Substituting 16 gauge shells in a 14 gauge gun is one quick and dirty example of emergency ammunition procurement. Cases thus fired will be ruined, which was probably one of the reasons why colonial governments in Africa and some places in South and Central America issued single-shot 14 gauge Martini shotguns to their native soldiers. It didn't take much gun to keep unarmed peasants in line, and the colonial overlords wanted to make it as tough as possible for unauthorized people to reload for the obscure 14 gauge guns.

"But I have never seen a 14 gauge gun!" sez you. You will if you work in such remote locations as rural Africa, and there is always the possibility something just as weird will come along. Nevertheless, it is still far more likely you will locate some 16 gauge rounds or empty hulls to reload than you will ready-made 14 gauge ammo. Sixteen gauge rounds are more common in some former European colonies and in Europe itself than they are in the United States. An estimated six percent of new shotguns sold in the U.S. are 16 gauge.

Ammunition

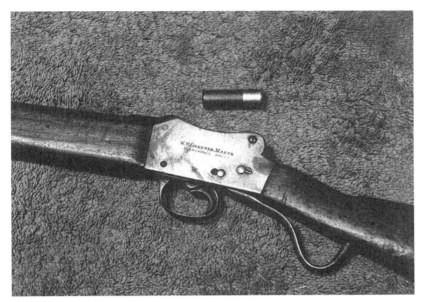

Sixteen-gauge rounds can be fired in this old Greener 14-gauge shotgun. When ammunition is scare, guerrilla gunsmiths must be alert for similar opportunities.

Firing .38 Special rounds in a .357 is another, even less exciting, example. I recently encountered a fellow in New Zealand who could not secure a steady supply of .357 brass for his revolver and replica rifle. As a result, he reloaded .38 Specials to .357 specs. Should these fall into the hands of someone who doesn't know about their potential, they may be in for a surprise, as the enhanced rounds will stress a regular .38 Special a bit.

In another notable instance, a fellow I knew once owned a pre-World War II Erma submachine gun, a Spanish rifle, and an Astra pistol, all of which could use 9mm Steyr rounds. These are about 9x23mm and are not common in some parts of the world. Regular 9mm Parabellum, i.e., standard 9mm Luger, is about 9x19 and will not cycle most guns made for Steyr rounds.

Left to right: factory 9x19; 9x23 made from .223 brass; .38 Special case which, with some work, can be made to work in a 9mm gun.

Standard .38 Special or .357 brass can be turned in to 9mm Steyr rounds, but the brass must be trimmed to .905 length, the rim lathe turned a few thousandths, and an extractor groove cut in the base of the case. All of this can easily be done by using an electric drill or drill press as a lathe and a small file as a cutting tool. For my friend this would have entailed a lot of fooling around, compounded by the fact that he didn't have a good, reliable source of .38 Special brass at the time.

What we did have was a pot full of once-fired, somewhat damaged military .223 brass. Using a small copper-tubing cutter, we lopped off the .223 brass neck and body 1 inch from the base of the round. We then carefully trimmed the body down to a length of .90 inches. These reloaded rounds worked like a top, but we sure cried when we lost any of them during practice ses-

Ammunition

Factory 9mm Steyr (right) next to a 9mm made from .223 brass.

sions. After having spent a lot of time on each individual round getting them "just right," it was a shame to have one fall among the rocks on into high grass where we couldn't find it.

Similarly, rounds for 9mm Makarov pistols can be made from .223 brass or from regular 9mm Luger rounds if they are more readily available. To do this, trim the brass back to right at .71 inches and load it like a .380 Auto round. Regular .380 ACP rounds will function reliably in many 9mm Makarovs. It's worth a try if you have access to this ammo. In that regard, it puzzles me to no end that it is considered tough to find ammunition for Makarov pistols. The procedures above are quick and easy. The only downside is that .223 brass may be slightly thicker walled, allowing for less powder in the load. So far I have not found this to be much of a consideration, much less a problem.

In yet another episode, we came upon an old, old South American copy of a Model 1911 .45 ACP, except the frame was made of softer steel and it was chambered for .38 ACP. Steyr rounds functioned very well in the pistol as long as we could get it to hang together. The problem in this case was that the magazine catch no longer caught properly. At embarrassing moments it would release, dropping the magazine. Other times it failed to hold the rounds in proper position to feed.

We cured this problem by brazing up the worn catch hole in the magazine and replacing the spring in the magazine catch on the pistol. Replacing the magazine with a new one would have been appropriate, but there were no new magazines anywhere close to *that* pistol!

Antique German 1879 service revolvers are still around. Most are likely in somebody's pistol collection, but in desperate situations they may have to be pressed into service. Use .44 Special cartridges from which about 20 percent of the factory-loaded powder has been removed. I *have* seen these old monsters fired with regular .44 Special rounds. It tests them mightily, but they do seem to function okay. If the pistol does let go, it may not hurt the user anyway (although the possibility of harm is *always* there). Unlike rifles and shotguns, which are held near a shooter's head, a pistol failure is seldom harmful to the user, even if it is very hard on the pistol itself. But again, we're talking last resort here. Don't do it unless you absolutely must.

Ancient 11mm French revolvers will function with .44 Special rounds too. Either remove half the factory load or reload the brass with dramatically reduced charges. There are probably more of these old 11mms hiding around former European colonies than the German varieties.

Japanese Arisaka rounds in 6.5 x 51 and 7.7 x 58 can both be made from U.S. .30-06 brass. This is good news because there are still lots of old, rough Japanese rifles floating around. Of course, .30-06 brass is probably the most common empty in the world after .22 rimfires.

Ammunition

Many ancient military revolvers are kicking around the world. Most will function with lightly loaded .44-caliber rounds.

In both cases, empty .30-06 brass must be worked into 6.5 x 7.7 dies, trimmed, and then fire-formed in the respective rifles. Forming dies can often be purchased from U.S. suppliers without raising suspicions, or they can be made in sufficiently skilled backcountry machine shops.

Substituting .303 British ammo in guns chambered for .30-40 Krag is another practice that really dates me. British .303 rounds are extremely common. Rifles for this cartridge were once scattered around virtually everyplace the British had colonies, which is to say everyplace.

Here in the U.S. of A., .30-40 Krag rifles came on the scene about 1892 or '93, went through several design changes, and then won some acclaim as our standard-issue smokeless powder repeating rifle during the Spanish American unpleasantness. Yet

Krags could not remain as our standard issue rifles because (a) .30-40 cartridges lacked power, (b) the rifles were expensive to manufacture, and (c) they could not be reloaded from stripper clips nor easily reloaded to top up partially empty magazines.

During its short life as America's military rifle, a considerable quantity of Krags was procured by the military. Gross numbers were not high, but relative to the size of our army and population at the time, quite a few were purchased. After 1906 when the 06 Springfield debuted (reconfigured from the .30-03), most of these Krags went on the civilian market. (That's how we did things back then.)

Given this dissemination of the rifle, it recently was not uncommon to run across a .30-40 Krag for which there was no ammunition, especially in Central and South America. In these cases, we simply scrounged up .303 British ammo and used it in our .30-40s. Worked just fine, but don't try subbing .30-40 into .303s. That doesn't work unless you reform the brass in loading dies.

There is one last suggestion. The addition of simple bushings to modify a cartridge should be considered too. In some cases this is a real winner. I have, for instance, fired 9mm rounds in .38 pistols by fitting the 9mms with thin steel washers in order to provide a suitable rim for them. Even primitive machine shops can turn out bushings that allow use of .308 (7.62 NATO) rounds in .30-06 rifles or .22 LR in .223 rifles.

Other than bolt-action rifles and single-shot shotguns, many common weapons won't cycle with such jury-rigged ammo, but like I said going in, maybe the best we can do is secure a device that can then be used to secure a real gun.

DETERMINING CALIBER

Most of the time it is fairly easy to determine a gun's cartridge requirements by knowing the make, model, and country of origin of the gun. At times, however, changes may have been

Ammunition

made, and frequently, way out in the field, there is no way to cross-reference the model against available literature. Our South American Llama—which looked identical to a Colt .45 ACP but was really a .38 ACP—is a good, real-life example.

Slugging the gun's barrel or, at times just eyeballing its size, will say something about its caliber, but not always. I ran into such an instance in southern Mexico on the Belize border.

A bandito had a venerable lever-action Marlin 97. It was another of these guns that in normal times would have been in somebody's collection except for the fact that the ammo tube had been replaced. I guessed it at about .38 to .40 caliber. The barrel was not marked, and there was no way to be sure about proper cartridges except by making a chamber cast.

Tremendous quantities of voodoo have evolved relative to doing chamber casts on firearms. There's no question that elemental sulfur and some low-melting-point bismuth compounds, as sold by professional gunsmith houses, give accurate, machinist-grade readings. But you as a guerrilla gunsmith ain't gonna have these materials to work with. What you *will* have is common paraffin, which is good enough for government work . . . or at least good enough to put together guns and ammunition with which to work on the government.

Proceed with making chamber casts as follows:

Clean the gun's chamber thoroughly. Apply a light coat of oil to the chamber and barrel throat. Hang or clamp the gun in a vise, muzzle down. Take great care to determine exactly how you will actually pour paraffin into the chamber.

Place a heavy wad of felt or cotton cloth on a foot-long piece of wire about as thick as a coat hanger. Cram this down into the barrel about an inch below (or ahead of) the chamber. Set it up so that the wire is sticking back out through the center of the breach.

Heat the paraffin till it is quite a bit hotter than would normally be required to melt it without setting it on fire. Be careful: paraffin can burn almost explosively. With the same propane

Chamber casts can be made easily with paraffin by inserting a barrel plug through the chamber on a piece of wire.

Ammunition

torch used to melt the paraffin, warm the gun barrel and chamber, playing the flame around the inside of the chamber.

Pour the hot, molten wax into the chamber in one big glop. Fill to the top where the bolt rests on the chamber. Unlike bismuth and sulfur compounds, paraffin is easily pulled from locking lugs should you get the casting too deep.

Allow about two minutes for the cast to set sufficiently. It may be necessary to refill the body of the cast as it cools and craters.

Place the gun with paraffin cast in a freezer or refrigerator for about four hours. After this chill time, remove the gun, give the wire a quick jerk, and out will come the complete chamber cast. Paraffin molds thus made will lack machinists' precision, but by comparing them to possible cartridges or by simply measuring, they will tell you what cartridge the gun is made for.

In our case, we discovered that our battered old Marlin rifle was a .32 caliber, not something in the .40 caliber range. It actually turned out to be an old .32-20 Winchester cartridge.

As a complete aside, we made ammunition for this gun by expanding the necks of .25-20 Winchester brass out to .32 caliber. This brass was then fired in the gun to blow out the cases to fit the chamber. Real gunsmiths call this process "fire forming." For bullets we used the heaviest .32 caliber pistol bullets we could find. These were somewhat undersized for that barrel, but they were lead. At modest ranges they worked just fine.

Our big problem involved the ancient .25-20 brass with which we had to work. Much of it was dingy and corroded. Half or more of the rounds we tried at first split before we were able to fire-form them. Annealing the old brass helped quite a bit.

Here is the method we used to anneal this pickup brass, much of which we found out in the hills.

Using a propane torch and holding the brass with gas-pipe pliers, heat the neck and the body of the cartridge until it shows signs of color. Keep the heat away from the base of the cartridge, which should not be softened. The pliers jaws gripping around the body will tend to keep heat up on the top half of the cartridge body.

Immediately drop the hot case into a pan of water to cool. Brass is different than steel—it softens as a result of quenching rather than hardening. It is hardened by much working (as in multiple reloadings) and by age.

Anyway, we got the old 97 going in sufficiently good form that the owners felt reasonably secure against government bullies.

In another notable instance, we ran across an old, battered Winchester 94. Again, this was south of the border. We suspected a .40 or .44 caliber cartridge that had been popular in the region, but the barrel was not marked. It perhaps was not the original barrel and was definitely a puzzle.

After chamber casts, we discovered it was actually a .38 caliber gun. It turned out to use a .38-55 Winchester cartridge. Not many of these kicking around any more!

We produced ammunition for this gun using old .30-30 Winchester brass that, in that place, was somewhat abundant. First we expanded the necks to .38 caliber. Then we blew out the cases to fit the chamber. We used the longest, heaviest .38 or .357 bullets we could find. These were cast lead bullets that, again, were a bit small for the barrel but seemed to work okay. Softer lead seemed to swedge up on discharge, filling the barrel snugly into the lands. It was another situation where we took a gun of unknown caliber and turned it into a usable tool.

PRIMER PROBLEMS

But what to do if the brass you encounter is Berdan primed? These are the European primers invented by an American that have detonator anvils as an integral part of the brass case rather than in the primer itself. In some places, Berdan primers for reloaders are becoming a bit easier to find. They are still not like American Boxer primers (invented by a European) in terms of availability and ease of reloading.

First you have to decap Berdan-primed brass. This can be a real chore or it can actually be done quite easily, depending how

Ammunition

Inside a Berdan-primed case showing anvil against which the firing pin crushes the primer.

you approach the problem. Cartridge cases made for Berdan primers have two or three small flash holes rather than a single, relatively large hole as found on Boxer-primed cases. A simple makeshift punch such as an ice pick won't take out spent Berdan primers without ruining the case.

Start the decapping process by scrounging or turning down a bolt or wooden dowel to just the size of the case mouth. Fill the case with water. Set it in a slightly opened set of bench vice jaws that will allow spent primers and water to vent out harmlessly below.

Align the plunger on the water in the empty case and give it a sharp smack with a heavy maul. I keep a pour pot full of water handy so the case can be topped up quickly and smacked again if the first try is unsuccessful. Pull off partially removed primers

Berdan case cut down to show small flash holes that frustrate decapping by conventional means.

with a needle nose pliers. Some have a bad tendency to hinge out halfway rather than expel completely.

Replacing Berdan primers is more of a struggle. They ain't like Boxer primers that can simply be pressed in to set. Set Berdan primers too deep and they will go off; set them too shallow and they won't detonate when struck.

Place the new primer in the primer pocket, finger tight. Turn the case over and gently push down on the same decapping plunger inside the cartridge case. If necessary, tap down on the plunger to drive the primer in flush with the base of the empty cartridge case. Even taking these precautions, you will likely get a premature detonation now and then. Eye protection and work gloves are recommended, and recap in a place where a little noise won't raise suspicions.

Ammunition

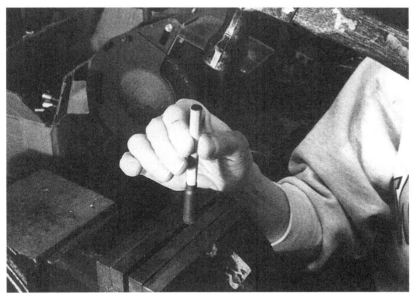

A carefully sized hardwood dowel used to punch a Berdan primer from a water-filled cartridge case.

It's another tedious process, but Berdan-primed cases can be converted to American Boxer primers if more of the latter type are available. I have performed this operation several times in my life when the circumstances suggested that it was a wise course. These circumstances must include the fact that you have a good supply of Berdan brass, lots of cheap labor, a source of Boxer primers, an acetylene torch, and small-diameter brazing rod and a #4 wire gauge drill (.2090 inch) for large rifle or pistol primers, or a #7 wire size drill (.2010 inch) for small rifle or pistol primers. Sorry about specking wire gauge sized drills, but they are the only drills that will work for this purpose.

Set the decapped cases upside down in a rack that is placed in a fairly large pan of water. The water should rise up the side

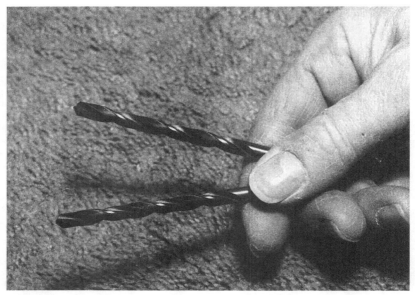

A number 4 wire gauge drill (top) produces a .2090 hole for large rifle or pistol primers. A number 7 wire gauge drill (bottom) produces a .2010 hole for small rifle or pistol primers.

of the cases about halfway. Heavily flux the empty Berdan primer pockets and, heating mostly the brazing rod, fill the pocket hole with molten braze.

Cooling will take a few seconds. Remove the case from the water and, using a bottoming drill (square-nosed), auger out a flat-bottom primer pocket for the new Boxer primers. Again, it will be a #4 wire gauge drill for large rifles and a #7 for small rifles or pistols. Depth of the pocket is .125, or 1/8, inch. This must be done accurately or the case will be overly weakened and/or the firing pin will not reach the primer or, in the case of a shallow pocket, will prematurely impinge on the primer. None of these outcomes are particularly good.

I drill primer pocket holes by centering the drill in a stationery holder on my drill press table and chucking the cartridge in the drill chuck. This way I turn the case rather than the drill. Seems a bit easier and more accurate this way.

Flash holes must now be drilled in the bottom of the primer pocket. Use a #50 wire gauge drill, producing a hole about .070 inch in diameter.

Obviously this is quite an undertaking. Expect to spend 10 to 15 minutes on each case. It's not something people would normally do if there were other alternatives.

SCROUNGING BULLETS

Finding correct bullets for some of these old cartridges can also turn into a real scavenger hunt. I have two suggestions other than the simple expedient of going to another size cartridge that may produce bullets that are close but not exact, the classic example of which is using heavy .38 Special bullets in a .38-55 Winchester or .45 pistol bullets in a .458 (which I frequently did when living in Africa).

All over the world, a tremendous number of bullets are made in custom molds. These molds are turned out by backcountry machine shops and profit from the fact that wheel weights are universal. Wheel weights make pretty good material for bullets when nothing else is available.

For jackets I have used everything from spent .22 cases to common soft copper tubing. One time I took some .38 Special bullets and soldered on a short section of .38 Special brass as a kind of bushing. This circle of brass brought the bullet up to .375 for use in a .375 H&H magnum. Accuracy suffered, and of course I couldn't load them very hot, but the round did work in the big gun.

In another instance in Somalia we needed some .457 bullets for a .470 bolt-action rifle. I found some .45 ACP bullets to which I soldered a ring of brass from some old .45 ACP cases to bring them up to .470. Again, this worked, but not like factory rounds. I am sure that the reverse could be done, that overlarge

A lead .45-caliber case with base removed creates a jacketed .470-caliber projectile. In an emergency, this is a good means of making usable bullets.

bullets could be turned or filed down a few calibers for use in smaller guns.

SALVAGING POWDER

Where can guerrilla gunsmiths get powder for cartridges? When and where it wasn't otherwise available, I have scrounged powder from dud artillery and mortar rounds, lost ammo we found but couldn't use, and even ammo we brought up from the bottom of a bay. In Rhodesia we purchased larger or different rounds from people who picked them up on battlegrounds. In Somalia and Kenya during the 1960s we still were able to acquire a huge amount of military surplus picked up from World War II North African battle sites!

Ammunition

Reading pressures from primers. Left to right: rounded primer indicating low pressure; primer squashed flat against the firing pin and bolt indicating high pressures; primer from regular factory load.

Successfully using this "off brand" stuff requires a great deal of cautious testing. Usually the powder is too slow in its present form, meaning it must be modified. Do so very carefully by grinding it up into smaller grains or flakes. In the case of artillery powder, breaking or slicing up the large chunks may be all that can be done. Again, results are about what one would expect in an emergency situation.

Mixing broken-up powder from an artillery round with powder from a .50 caliber may produce something usable, though this is a potentially risky endeavor that is best left for extreme emergencies. Caution is required to maintain consistent proportions one week to the next. Start off on the safe side by reloading with obvious underloads.

While a number of field-expedients may have to be utilized in an emergency, it is always best to use powder from scrounged rounds that are similar to the rounds being reloaded (i.e., large rifle for large rifle, small rifle for small rifle, pistol for pistol). Swapping and mixing powders is always a cautionary task, even when the rounds appear to be very similar.

How to know if you are getting into trouble with excess pressures in your reloads? By the ancient and honorable art of reading primers of fired rounds. I do not want to bore readers with stuff they already know, but I haven't seen mention of primer reading in recent gun literature. It definitely is something you should know if you don't already understand how it is done. I can't recall who taught me or how I learned. It was many, many years ago.

Reading primers works on the assumption that dramatic overloads will first reveal themselves by blowing primers out of their pockets. As a subset of that condition, modest overloads will flatten primers against the bolt of the gun. Soft primer metal will reveal this flattening by taking up a squashed, almost smeared look. These primers will mirror tiny imperfections in the bolt and even mill marks on the firing pin itself. At the other end of the spectrum, light, wimpy loads don't even take the roundness out of a primer when the cartridge is fired.

With a bit of practice comparing two extremes in front of you along with a factory load, it is possible to get a pretty good idea of the internal ballistics of your rounds. Look very closely with a magnifying glass, if necessary, to see the condition of the primer. If the firing pin strike is deep and round, there is little likelihood that pressures are becoming excessive. See the photo on page 81 for an example of the extremes.

All of this should get guerrilla gunsmiths to at least look at various alternatives to accommodate their individual ammo shortage situations. Creativity in these cases really does count 99 points.

chapter 6

Subcaliber Adapters

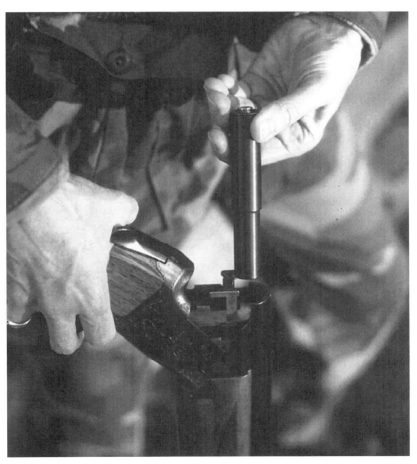

Inserting a 12-gauge adapter in a double shotgun.

Guerrilla Gunsmithing

FAITHFUL READERS ALREADY KNOW that subcaliber adapter devices are one of my favorite topics. This is because I have seen them put to so much good use around the world.

I am impressed with these little gizmos for several reasons:

1. They allow users to deploy otherwise abundant ammunition of an entirely different caliber and type in guns that normally would be of only limited use. Without available ammo, these guns are, of course, unusable, unless one wants to orchestrate a psychological show of force. During World War II, for instance, Norwegian resistance fighters sometimes took photos of dozens of men with what appeared to be working weapons, giving the impression of a large force so they would appear to be more credible in the eyes of their British suppliers. This is the opposite of what we want to do with subcaliber adapters.
2. Subcaliber adapters allow use of old guns that are so worn out they have little purpose outside of being holders for adapters. Hostile government forces sometimes will allow ownership of these types of weapons.
3. Authorities seldom seem to recognize neither the nature nor insidiousness of these little devices. They basically ignore them, even when seen laying out on a desk or shelf someplace.

I have personally carried adapters all over the world without anyone even recognizing them for what they were. I even took a 12 gauge/9mm model into gun-paranoid Singapore, where private ownership of firearms can mean life-imprisonment or even capital punishment.

These devices are incredibly inexpensive to purchase. MCA Sports/Ace Bullet Co., 2800 W. 33rd Road, Anchorage, Alaska,

Subcaliber Adapters

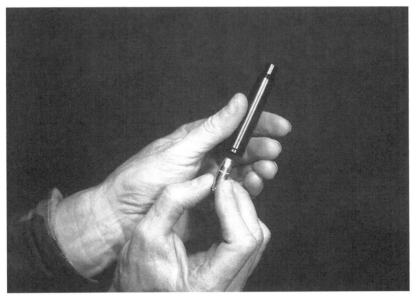

Loading a .300 Win Mag adapter with a .32 ACP round.

99517, stocks many types or can make them on a custom basis very quickly. Some models allow .30 carbine to be fired in .308, .30-06, or 12 gauge guns. Others fire 7.62x39mm in 12 gauge; .30 Luger, .30 Mauser, or .32 ACP in any .30 caliber gun such as .300 Win Mag, .303 British, .300 Savage, or .30-40 Krag; and on and on and on. There are virtually no limits once you are in the mood to think about the concept. If nothing else, I would immediately order one of MCA's catalogs to see what's commercially available today.

In this day and age, many adapters are configured to fire .308s in .30-06 rifles. Surplus dealers commonly carry these devices, which were originally made by foreign governments that wanted to convert their existing guns to .308 NATO. Cost per adapter is, at most, $5. Sometimes .308 adapters are left

Inserting a loaded .308 adapter in an FN assault rifle.

permanently in chambers of .30-06 rifles. Autos won't reliably cycle, but all other guns will work just like they should with these units in place.

TYPES OF ADAPTERS

Adapters fall into two different categories. The first type, called a chamber adapter, makes use of a cartridge of the same caliber as the gun itself, using the gun's barrel to stabilize the projectile from a different, smaller round. Bullets fired are, in this case, the same as the bore of the rifle. A good example is .22 LR rounds in .223, .22-250, or .22 Hornet. Most of the cartridge conversions listed above are of this class of adapter.

Subcaliber Adapters

One type of subcaliber device makes use of the gun's existing barrel to stabilize smaller pistol rounds that are also the same caliber as the gun. Shown from left to right are a .300 Win Mag, .30-06, .308, and .30-30, all of which accurately fire a .32 ACP round. On the far right is a .223 insert that fires a .22 LR.

The second type, the subcaliber adapter, has its own rifled barrel as an integral part of the device. This is how 9mm rounds can be safely and accurately fired in 12 gauge shotguns.

As a result of having their own rifled barrels, these devices can produce some strange configurations. We owned a .22 LR insert as kids, allowing us to fire .22s in our 12 gauge shotgun. The bore on the adapter was offset to handle rimfires being detonated by a centerfire gun. I have also had inserts made to fire .22 LRs in a .300 Win Mag. This is a very handy device for potting grouse for the pan around hunting camps.

Getting .22 rimfires to work in any centerfire gun is a challenge. Modern devices as made by the folks at MCA Sports/Ace Bullet Co. make use of a small firing block that has a ridge milled across it. When a centerfire firing pin strikes the block, its ridge

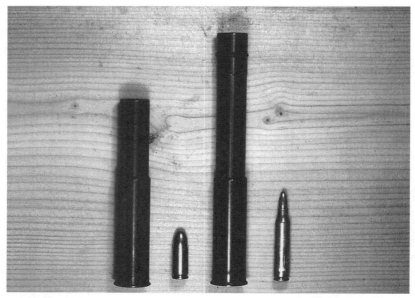

A second type of subcaliber device has its own rifled barrel, allowing use of such rounds as 9mm or .223 in a 12-gauge shotgun. Shown here are adapters for 9mm (left) and .223.

creases the rimfire round, causing ignition. See the photos for a more clear explanation of this process.

USING ADAPTERS

Subcaliber adapters can be very quiet in action while still maintaining a surprising amount of power and accuracy. My daughter, for instance, uses a 9mm device in her 12 gauge single-shot shotgun to hunt deer in suburbia. She gets about a 1-foot group at 150 yards. At that range, a 9mm will punch through a deer's ribs and lungs.

Loading and reloading, on the other hand, is slow to downright tedious. They work best in bolt-action and single-shot

Subcaliber Adapters

Striker assembly used to discharge .22 LR rounds in a .338 Win Mag subcaliber device. Note the ridge on the breech block that is struck by the rifle's firing pin.

shotguns, but models are out there for semiautos, pumps, and lever actions if these are the firearms you must deal with. In these cases, not only must the device be individually inserted and then withdrawn from the gun, empty cartridge cases must be picked from the adapter.

Firing speed can be marginally enhanced by providing more than one device to each shooter. Double devices in double guns and all of that. It's better than stones, knives, and bows and arrows, but at times one may wonder how much better.

If all of this seems unduly confusing, look at the photos. It really is simple, straightforward technology. That there are two types of adapters should be clear, as well as how they are assembled and how they are used.

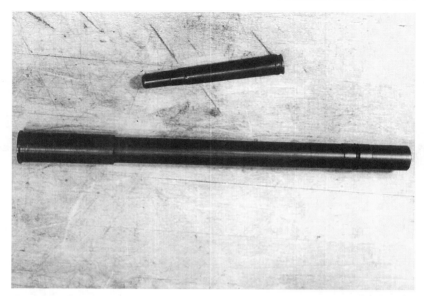

Two subcaliber devices with their own rifled barrels. At top is a .22 rimfire for .338 Win Mag; bottom is a long-version 7.62x39mm for 12-gauge singles or doubles. Longer devices are more accurate, but they are also more difficult to deploy.

BUILDING AN ADAPTER

Subcaliber adapters are relatively easy to manufacture in backcountry machine shops. This is especially true of versions that call for simple, straight-case pistol rounds to be fired in shotguns or rifles. Good examples include .32 ACP in .308 or .30-06, .22 LR in .223, or even 9mm in a 12 gauge. The job can even be accomplished with aluminum or brass bar stock if no provision is made to rifle the barrel. Much of the work can be done on a drill press if a lathe is not available.

As an example, to build an adapter that will fire 9mm or .38 Special in a 12 gauge shotgun, start by cutting a piece of bar stock down to 12 gauge dimensions. Model it on a 12 gauge empty from the gun that will be used with the adapter. The

Subcaliber Adapters

finished device can be a bit longer than a regular round, extending the adapter's barrel down into the barrel of the gun a bit.

Drill a standard 23/64 inch hole down through the exact center of the turned-down bar stock. This becomes the barrel. Size of the barrel is .359 inch, and it will work with either 9mm or .38 Special rounds. There will even be a bit of accuracy with soft lead bullets. Careful work on a good quality drill press is usually sufficient for this task.

Standard .38 Special cartridges are a few thousandths smaller in the body than 9mms. Chambers that will accept both can be drilled with a standard 25/64 inch drill bit. For .38s, the chamber will be a bit large, but it beats trying to find a particular reamer for either cartridge.

Use a bottoming drill to auger down till the cartridge rests in the device at the proper depth. Headspace is controlled by the depth of the drill hole. Don't go too deep or the gun's firing pin won't reach the cartridge primer.

This, then, is yet one more way to keep guns operational using off-standard, battlefield pick-up surplus, or antique guns and a strange assortment of ammo.

chapter 7

Fiberglass

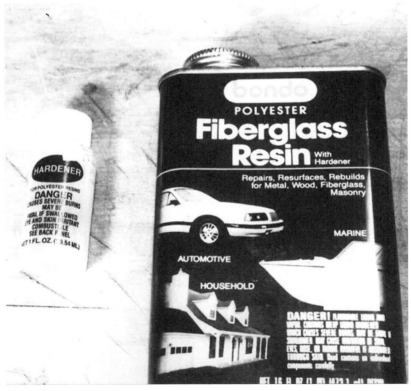

Common fiberglass resin and hardener—the guerrilla gunsmith's friend the world over.

Guerrilla Gunsmithing

THERE IS NO BETTER MATERIAL with which to hold sick and ailing guns together than fiberglass, especially if it doesn't matter what the end result looks like. I am continually reaching for cans of resin, hardener, and fiberglass patching cloth when working on guns in tough places. Fiberglass allows you to repair guns that may not otherwise be repairable. It is also an incredible aid to accuracy, as you will see.

Most Americans in the gun culture already know about the role of snipers in settling issues involving government agents. What is generally not known is the fact that ranges over which snipers fire are often modest. Currently, common soldiers only reliably produce casualties at ranges of 300 yards or less. Police sniper shots rarely exceed 70 yards. Highly trained military snipers can engage targets in excess of 1,000 yards, but that is an entirely different scenario with which guerrilla gunsmiths will not have to deal. Therefore, rifles in the hands of guerrillas should shoot accurately over typical engagement distances. Sadly many do not.

I was living in Cuba in late 1958 when Fidel Castro made his final move to power. It all started in the Sierra Madre Mountains with an attack on a government outpost. This attack was reported to me at the time as being the final action calculated to put Fidel in Havana by Christmas. These relatively minor events would not have successfully rolled up the Batistas had Fulgencio not been such an incredible weenie. Upon hearing of the insurgents' initial, but slight, success, Batista fled the country.

But this avoids the real issue relative to this account. According to contemporary reports, Fidel's attack on the Batista outpost was successful only because he personally managed a one-shot kill on the army's radioman in the opening moments of the action. I was told that this was done with a scoped Winchester Model 70 in .30-06 caliber at the incredibly long range of 175 yards!

Even allowing for victor's exaggeration, this was really not much of a shot, but it did manage to scare away the

Fiberglass

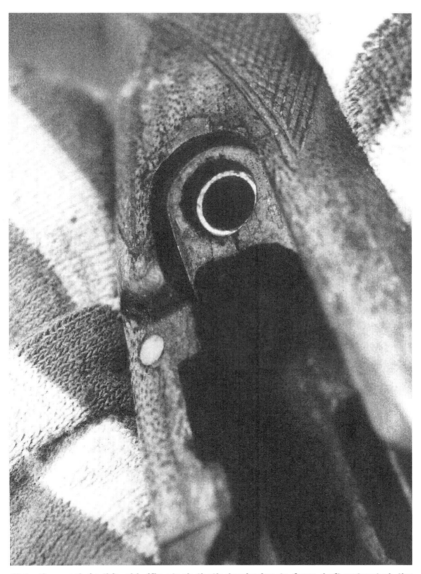

Note the crack in this old rifle stock that's beginning to form. Left untreated, the stock will soon be destroyed.

Recoil lug on an old rifle that has been fiberglassed to preserve the stock and increase accuracy.

Batista government. Now, for 40 years we have had Fidel at our doorstep.

The point is that even modestly long-range shots can and have changed the course of history.

LOCATING SUPPLIES

To find fiberglass supplies, you are going to have to go to a body and fender repair shop ("panel beaters" they call them in many places around the world), a boat builder, or a factory that uses fiberglass as part of its manufacturing process. Look for fiberglass resin, cloth, and a bit of hardener as well as some release agent. Hardener can be difficult to find in any quantity in the Third World, but in this day and age I have been able to purchase

Fiberglass

Common paste wax makes a good, inexpensive release agent for fiberglass, but try specific kinds first to make sure they work.

fiberglass supplies and accompanying thinners in remote areas of South Korea and Indonesia. I have even found it in Burma, Cuba, and Algeria, where it was used to repair boats.

The second problem is finding enough good, reliable release agent. Most gunsmiths don't want stocks melded to barrel and action or screws frozen in their holes. Therefore having enough quality release agent is a necessity.

Fortunately, there are some quick improvised remedies for guns that you do not want to become hopelessly locked in fiberglass. Johnson's floor wax will work very nicely if you can find it in the country in which you are operating. If not Johnson's, try any floor wax you can find on the shelf. In many cultures floor waxing is uncommon, but if the stuff is available, it is a cheap, easy, and effective way to keep fiberglass from sticking.

Coat some trial gun parts three or four times with the wax. Mix up and apply a tiny batch of glass and see if it will release after it is dead hard. If it doesn't release, no harm is done to these scrap pieces.

If and when you have a batch of fiberglass that has hardened in the wrong place, don't despair. No solvent on Earth will cut hardened glass, but temperatures of about 300°F will destroy it. This is not hot enough to damage most gun parts, including stocks. I learned this trick removing scope mounts and front sights that had been temporarily epoxied to shotguns. Fiberglass isn't epoxy, but other than some special factory-cured epoxy resins, both will release under heat well under temperatures necessary to silver solder or anneal steel.

So both will release under heat well under temperatures necessary to silver solder or anneal steel.

Some guerrilla gunsmiths use Scotch tape on metal parts to keep glass from bonding to them. I don't care for this procedure, but I do use modeling clay to block off places on stocks I am bedding where glass is not needed.

As an aside, many gunsmiths don't like automotive fiberglass resins because they (the resins) are too thick and because they are usually colored a dull, dead gray. Guns, these fellows claim, look much better in brown, even if they are only going to be used as guerrilla weapons.

Other than painting after it is dried, thick, heavily pigmented automotive resin is tough to color. Automotive resins can be thinned a bit with lacquer thinner or acetone, also available at body shops, hardware, or paint stores.

GLASS BEDDING

There is no quicker way to improve accuracy dramatically in many beat-up civilian guns than to glass-bed them. Glass bedding has worked marginally well for me when patching up old military rifles and shotguns, but its real charm becomes apparent

Fiberglass

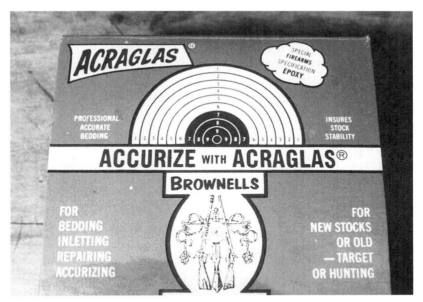

Brownells makes a prepackaged fiberglass kit that is very handy for guerrilla gunsmiths. In an emergency, the same basic materials can be found at automotive and boat shops worldwide.

when busted and dinged commercial weapons are treated. This is because most commercial guns are far more fragile than even the worst Italian or French military small arms. Commercial weapons quickly fall apart under the stress of the heavy service requirements of guerrilla actions, but fiberglass can bring them back to life.

I have worked on cheap, commercial export .303s, .308s, and a 7.5mm French rifle that wouldn't hold a 10-inch group at 100 yards. They shot 4-inch groups after bedding. I've wondered if locals who may be unaccustomed to guns and shooting would be capable of this much accuracy, but the ones I've worked with were happy with the bedded guns, so I was happy.

Professional bedding kits are available from Brownells, 200 Front Street, Montezuma, IA, 50171, for those who wish to

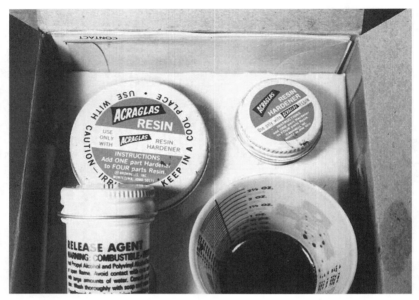
Individual components of a Brownells fiberglass kit.

accomplish this the quick, easy way. These inexpensive kits are extremely simple to use but lack versatility. You must, for instance, use most all of the supplies on one little job. And besides, these kits will not be available to gunsmiths laboring in remote areas under guerrilla conditions.

When glassing a rifle either with a kit or your own acquired supplies, start by removing the action from the stock. Completely coat the barrel, action, and trigger guard assemblies with four coats of release agent.

Take a close look at the stock to determine if it is being beat to pieces on the recoil lugs (found just behind the trigger opening on the stock and sometimes in front of the magazine well). Powdered, mashed wood or fiberglass—indicating much use—as well as overfrequent, overgenerous oilings lead

Fiberglass

to deterioration of the wood. Continued use without repair can lead to a splintered, split stock. Fiberglass applied to these areas will slow the process, repair the stock, and result in better accuracy.

Clean the damaged wood from the recoil lug area of the gun. Use a rasp or chisel to remove and roughen wood under the barrel all the way out to the end of the stock.

Place five drops of hardener catalyst in one liquid ounce of resin, if the temperature is about 65°F. When temperatures are lower, add an extra drop or two; if it's warmer, cut back to three to four drops. Quickly and thoroughly mix the stuff up. Use as little of the precious catalyst as necessary. More catalyst equals greater shrinkage of the final set when the fiberglass hardens. We want to end up with a free-floating barrel but not one with the 1/16 inch clearance or more caused by excessive shrinkage.

Barrels and actions coated with floor wax release agent are perfect for bedding. The wax is very thin, creating a hairline (or less), free-floating arrangement after the wax is removed. After everything has set up, remove the floor wax by washing the pieces thoroughly in acetone or lacquer thinner.

After placing the treated fiberglass resin in the stock back around the recoil lug up through the end of the wood, tighten the barrel and action into the stock with the regular stock screws. Be certain the fiberglass is everyplace you want it, especially if the stock is being repaired in the same operation. Even wood with some old gun oil in it can be successfully glassed if internal surfaces are roughed up a bit and you work the glass into all of the cracks (or you can purge the oil out of the stock as outlined in Chapter 10).

Be sure all stock screws are liberally coated with release agent. Screwing the stock into the action will cause excess glass to ooze up around the barrel and trigger guard. Best to deal with this excess before it hardens rather than after. Using dry cloths, clean as much of it away as possible. Then, using cloth dampened with acetone, finish cleaning up.

Guerrilla Gunsmithing

If you want to leave it, excess fiberglass that hardens can be sanded or filed away. (This isn't finish gunsmithing by anybody's standard.) Fiberglass that has set up is fairly soft, but it is easier to clean, I think, while still a liquid. Acetone will cut the liquid glass till about three minutes before it really commences to harden. But be careful: acetone washes away floor wax release agent. It's easy to create a mess with the stuff.

No matter what, I leave my fiberglass to cure at least overnight. Even with proper release agents in place, removing the barreled action from the newly fiberglassed stock can be a trick. After removing all retainer screws again, give the barrel a sharp crack with a rawhide or rubber hammer. This usually snaps it out of its form. As suggested, clean the wax from the metal parts, creating what is probably a 1/64 inch free-floating barrel.

Fiberglassing a gun is really not high tech at this level. After locating supplies of fiberglass resin, hardener, cloth, and release agent, it's pretty much downhill. Take your time and experiment on noncritical pieces of scrap before going for the brass ring. With a little practice, it can be a cure for lots and lots of guerrilla gunsmithing ills.

chapter 8

Guns and Glue

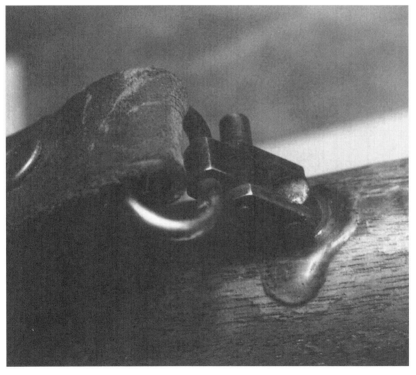

A sling swivel repaired with epoxy.

LIKE FIBERGLASS, MODERN TWO-PART GLUES can have a material impact on one's ability to work on guns under less-than-ideal circumstances. At times when I had nothing else or was in a big hurry, I have successfully used epoxy glue for everything from attaching sights to repairing smashed stocks, securing slings (without posts or swivels), mounting scopes, repairing magazines and some worn internal parts, saving stripped threads, and mounting new cheek pieces. In one notable case, I rebuilt an AR-15 sear using Devcon B compound by filing the sear back down to proper dimensions after rebuilding it with this epoxy material. (I'd be wary of habitually rebuilding critical parts this way, but in a guerrilla gunsmithing situation you may not have any other options.) In another, I glued front and rear sights on a 12 gauge pump shotgun that was subsequently used over a period of four years in four different states, Canada, and Mexico. As far as I know, those sights are still on that gun. The current owners probably don't even know the sights are glued rather than soldered.

Modern epoxies are both similar and different than fiberglass. Both result from the chemical merger of two components. The difference is that fiberglass catalysts create their own internal heat in resin, which then solidifies irrevocably, whereas epoxies actually involve the merging of two different sets of molecules into a final compound.

Tensile strength of some modern glues runs over 4,000 lbs. per square inch! In some cases this is as much, or significantly more, than solders and brazes traditionally used to mount sights.

Common wisdom states that unless a glue can dissolve and/or penetrate a surface, it cannot hold the join with certainty. This is why some wood glues produce unions that are virtually stronger than the parent material, while some epoxies used to join steel sometimes produce mediocre results.

WORKING WITH GLUE

Gluing may not always be easier or preferable over soldering or brazing. It's just that doing this work may be more possible

Guns and Glue

Super glue (left) is insufficiently strong for use on guns. Regular old slow-acting epoxy (right) is much more efficient than the five-minute variety.

with materials on hand or those that can be readily procured, and glue is often easier to get hold of than soldering equipment. There are, however, a few cautions to keep in mind if one is to use glues to do gunsmith work successfully.

First and foremost, not all epoxies are created equal. For instance, so-called fast-acting "five minute epoxies" are only about half as effective (i.e., have half the bonding strength) of longer setting, slower types. Don't try to use these quick-drying epoxies for gunsmithing work. A good rule of thumb is to purchase slow-acting epoxies, use them only in surroundings of 75° F, and allow at least 36 hours to cure.

After a *thorough* cleaning, roughing surfaces that are to be glued with a coarse file, old drill bit, sandpaper, or whatever helps immeasurably. Gouging deep ruts in sight bases and on

Devcon B can be used to build up, repair, and in some cases actually construct gun parts.

Guns and Glue

barrel surfaces is also effective. I have even used a metal punch to stipple areas to be glued (if you try this, create a bell-shaped stipple if possible). Glue will hold to such surfaces much better than it will to a glass-smooth, polished surface.

Mix two-part glues thoroughly, but work quickly. I don't like to be stirring them after they start to thicken and set up. Get all parts coated, wait until excess epoxy in the cup starts to thicken, then clamp the parts together securely and leave them until the stuff cures. Monitor the cup with excess epoxy in it as a guide to how the hardening is progressing.

Common, white glue such as Elmer's will work well on broken stocks if the break is a clean, new one and the stock has not been heavily treated with oil. Alas, most old military rifle stocks are absolutely saturated with oil. Epoxies, on the other hand, do a pretty good job of mending these oily breaks. As a result, I use nothing but epoxy to mend stocks. I know about all of the tricks to draw oil out of a stock, but the bottom line is epoxies hold better under adverse circumstances, so why even take a chance with white glue?

In 1965 a pen of 200 lb. market hogs burst out into a corral area where I kept a Marlin 39A .22 rifle. I kept the rifle at hand to shoot disease-bearing sparrows, but the hogs had a happy three hours rampaging over it, pushing it back and forth on a rough concrete floor, walking on it, defecating on it, and generally tearing the gun up. In the end, the stock was in three pieces and the scope was heavily gouged by the concrete. Now look at the photo of the repaired rifle on page 127. I doubt if you will be able to detect the places where epoxy mends were made. The scope is a replacement, but I still use the rifle extensively.

TYPES OF GLUE

What kinds of epoxy work best? Devcon B is a heavy steel-filled epoxy that I have used to mend torn magazines on Enfield rifles. Smear it on smooth, allow a couple of days to cure, polish

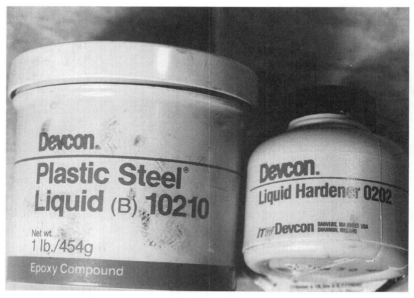

Component parts of versatile Devcon plastic steel.

sufficiently to allow internal springs to work, insert in the gun, and away you go.

Brownells sells a vent rib and poly choke adhesive that is also very nice. It is flexible when cured and bonds metal with great strength. But this information is of absolutely no use to guerrilla gunsmiths, who are likely to be operating in places that never heard of Brownells, much less Montezuma, Iowa, where the company is located.

In the Third World or under rough conditions in the United States, we are going to have to use whatever epoxies we can find off the shelf. The problem here is that the maximum shelf life of epoxies is only about four years. In many places in this world you stand an excellent chance of getting hopelessly stale materials.

Guns and Glue

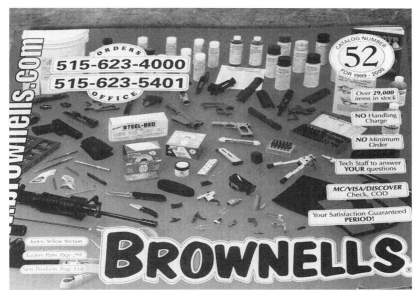

Brownells publishes a catalog crammed with tools and supplies of great value to the guerrilla gunsmith. It would be wise to stock up on crucial items now.

So how does one acquire fresh, workable glue for gunsmithing under tough conditions?

Run, don't walk, to the nearest shop in the area that does car and truck window installation. They will have urethane glues that, for the most part, will be fairly fresh. It may not be as strong as some epoxies, but it bonds to steel and is generally available. Beg, borrow, or buy a small quantity of this stuff from them. As with all adhesives, first try it on scrap to see if it will work properly. Chances are good it will. Anywhere in the world, this is the single best source of fresh, properly compounded glue sufficient for gunsmithing purposes.

Again, don't even attempt to use single-component quick-drying glues. These hold very, very poorly on metal, don't penetrate well, and do not have sufficient body strengths.

Silicon grease or our old stand-by, Johnson's liquid floor wax, makes a good epoxy release agent when such is required. I haven't yet needed a release agent when working with epoxy, but I might sometime in the future. That is not to say I haven't *wanted* to remove a rear sight or scope base that I had epoxied in place.

When removal is in order, heat the bonded parts slowly with a propane torch or even in a camp fire to about 300°F. Like fiberglass, epoxy turns to mush when heated. Released parts can be cleaned with a knife blade or scraper.

One more technology comes to mind under the heading of glues and gluing. It relates to the fact that good old Loc-Tite makes a product that restores stripped threads. It is my understanding that their product, "Form-a-Thread," will hold to about 125 lbs. of turning torque, which is actually quite a bit for stripped threads. Usually I brush such screws with fine steel wool to make them grab again in a stripped-out screw hole. This is okay on larger bolts, but on tiny thread holes it is much better to use Form-a-Thread if you can afford it. Less than two ounces costs $22 in the United States; it goes up from there overseas. It can be purchased in automotive shops, and most auto repair shops, even in Africa, seem to have Form-a-Thread. Stripped and worn threads will be one of the major repairs you are forced to deal with in a guerrilla gunsmith situation.

chapter 9
Clips and Magazines

Old pistols of this type are practically worthless when their magazines are gone.

My personal library contains a fairly large selection of volumes on the subject of gunsmithing. Most of these books were accumulated during the days of my youth in the 1940s and '50s. Significantly, not a single one of mentions procedures relative to faulty, worn, and dinged magazines. There isn't even a reference to magazines in the table of contents.

Strange, because at that time our country was emerging from a war that produced a truly large number of new weapons designs featuring detachable magazines. Even before World War II, we had the likes of the Colt 1911 .45 and 9mm Browning Hi-Power pistol, both of which functioned with magazines.

My personal experience suggests that magazines are one of the first components of a weapon to wear and break. "That's why magazines are issued in such large quantities," thought I. But to be sure, I checked with three gunsmiths, two military guys with combat experience, and three generic gun nuts.

Their assessment was unanimous: magazines wear out and break quickly and frequently. It's the cause of 70 percent of modern weapon failures, they figured. It's also one of the reasons why magazines are issued with weapons at ratios of 20 to 1 or more. My panel of experts also suggested that if a weapon is not functioning properly, you should first look at the condition of the magazine. Is it presenting the round properly so that the bolt can pick it up? Failure to place the cartridge can be a problem in virtually any weapon. I have even seen it happen with old Enfields and, of course, Krags. In many cases the main magazine spring is worn out, the follower is grabby, raspy, or perhaps misshapen, or the magazine catch notch is worn and isn't holding the magazine properly in the gun.

WEAK SPRINGS

Starting from the top, when magazine springs get old, wimpy, and weak, they won't push rounds up high enough or at the correct angle to allow pickup by the bolt. It isn't likely, but

Clips and Magazines

Feed ramps such as this one on an obsolete Grendel pistol are subject to damage and problems. Check the magazine follower first, then the feed ramp, if the gun is jamming on feeding.

it could be that the cartridge feed ramp is scratched, nicked, or otherwise unworkable. Check the feed ramp to be sure, but chances are it will be a weak spring inside the magazine that's causing the problem.

In a desperate situation, a replacement spring probably won't be available unless you happen to have common magazine parts in your bone pile. Alternately, new springs can be custom made by wrapping a piece of spring steel wire around a wooden or steel mandrel. The hardest part of this job is getting the mandrel. If I can't find a ready-cut piece of steel that will work, I make forming mandrels out of a piece of hardwood block that's been cut and shaped the same size as the spring. Thankfully, I usually have old, weak springs around to act as templates. Wrap the spring steel wire around the mandrel to

Guerrilla Gunsmithing

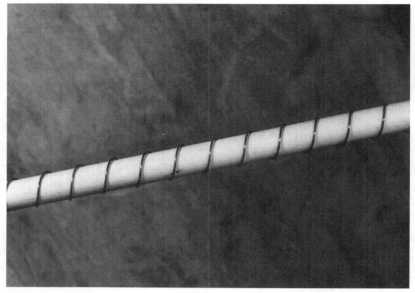

Spring steel wire can be wound on a suitably sized mandrel to make a coil spring. Having parts of the old spring as a pattern for the new one is very helpful.

form a new spring. Smack it with a hammer on the corners as you wrap. Make as many coils as the original, assuming the same diameter spring wire.

Finding proper spring steel wire can be a real challenge. In this country we buy the stuff at hobby shops or order it out of catalogs but out in the Third World it just isn't available—there are no hobby shops in places like Somalia or Zambia. I have looked in junkyards, automotive supply stores, cabinet and furniture shops, and manufacturing plants of all kinds and still come up short. One time in Indonesia I was able to find some usable wire at an oil-drilling site. Places I haven't tried but which may produce something include well drillers, aircraft mechanics, and even orthodontists offices if there are any around.

Clips and Magazines

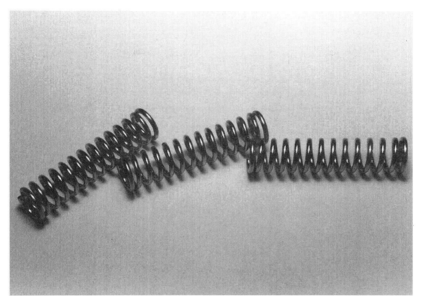

Coil springs used to strengthen an Enfield magazine.

It is very easy to spend a couple of days searching as I did, only to come up with absolutely nothing except some insignificant bits and scraps. In that case we were not able to manufacture our own magazine springs, so it was time for plan two: affixing stock coil spring in magazines as an auxiliary source of force.

To do this, glue two short, stubby posts top and bottom in the magazine and slip a long coil spring on these posts. Two coil springs may be required to get the follower to work properly. This procedure may also entail cutting a small opening in the bottom of the magazine housing to get the spring or springs in place. We did this with an M-14 magazine in Thailand. The Chinese fellow who did the work was, in my opinion, incredibly skilled. If all this seems confusing, look at the diagram illustration. Even weak coil

Guerrilla Gunsmithing

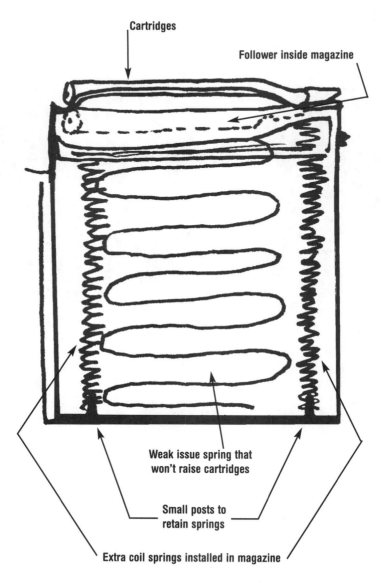

Cutaway view of a rifle magazine with helper springs installed.

springs will beef up faulty magazine springs at a time and place when nothing else is available.

FAULTY FOLLOWERS

If your magazine is malfunctioning, the second problem to look for is grabby, misshapen followers. I don't know how these get bent, broken, and scarred, but they frequently do. Check them out thoroughly to be sure the followers move up and down flawlessly, with an almost fluid motion.

If the follower glitches, polish it and the magazine housing with a very fine file or steel wool, or lubricate with STP or something equally slick. (As with all lubricants, be careful of grit and dirt mucking up the works.) Take care not to use coarse, rough tools or abrasives on these follower pieces lest you make them worse than they already are.

Magazine housings can become stippled, bent, or dinged. Disassemble these magazines, place a piece of steel or hardwood block inside the housing, and pound the problem areas back into shape against it. Use a brass hammer if one is available, or else work on the metal relatively gently with a steel hammer.

Longer stick magazines from submachine guns and assault rifles have a tendency to become bent or kinked after rough service. Many times these are simply thrown away, only to be picked up by people who want to sell them to us. In most cases it would be better to ignore these used magazines and start new, but guerrilla gunsmiths are unlikely to have a reliable source for new magazines, so it may come down to a matter of repairing and refurbishing the old, no matter how badly they are damaged.

After straightening and smoothing badly damaged magazine bodies, fill in any tears or holes with Devcon B epoxy, fiberglass, or braze. In these cases, glues and epoxies are at least as good as brazing, which, if not done skillfully and quickly, will hopelessly damage thin-steel magazine bodies.

Check magazine followers very carefully. They should operate with oily precision.

WORN CATCH NOTCHES

The third common problem with magazines is worn catch notches. Dealing with them in older magazines can be a real challenge.

I currently have at least eight rare magazines that go with specific military semiautos. All are worn beyond practical use. I was able to purchase (at times at great expense) newer replacement mags for these guns, yet the original magazines are still valuable, and I am reluctant to throw them away. As a result, my heirs will eventually come upon a whole drawer full of magazines that are too worn to use and too valuable to throw away. No doubt someone, someday, will try to rebuild them.

In most cases these magazines can be revived by carefully rebuilding and recutting catch notches on them. This is not an

Clips and Magazines

An otherwise worthless FN magazine repaired with brazing.

easy or quick procedure, but those who have no other magazines may be forced by circumstance to fill in old, worn catch notches with braze or Devcon B and file out a new notch.

I have worked an entire day at rebuilding a single damaged magazine. It often helps immeasurably to take the grips off a pistol to see more clearly where the catch notch should be. In the case of a submachine gun or assault rifle, look up into the magazine well to get an idea of how things should work. Do this before doing any work on the magazine relative to worn catch notches.

BENT LIPS

Dealing with worn, frayed, and bent magazine lips on old magazines is another common challenge. Ideally you will have

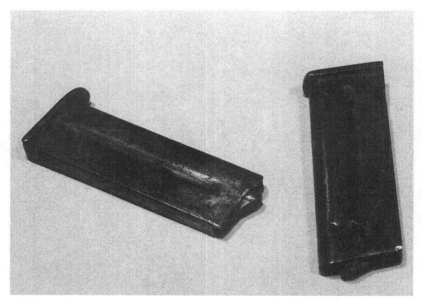

Old Astra pistol magazines—too worn to use, too valuable to throw away.

a relatively new magazine handy to compare to. If not, place the magazine in the gun, cycle the bolt and see what it looks and acts like.

Assuming the magazine is being held correctly in the gun, look to see if the magazine lips hold the cartridges properly for pickup. If not, are the lips bent and worn to the point where they won't hold cartridges properly? I have seen a number of Vietnam-era magazines in this condition that have been rebuilt by application of brazing on the lips. In one case, somebody welded in a completely new set of metal strips on the top of the magazine. (I believe a watchmaker or perhaps a jewelry expert did that work. It was very intricate, far beyond anything I could do. But it is important to know that this type of work can actually be done.)

Clips and Magazines

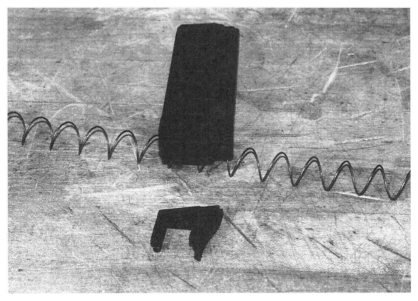

Gun magazines are usually comprised of four components: a body, spring, follower, and bottom access door.

BUILDING FROM SCRATCH

In some cases it may work to use old magazine bodies as models to make an entirely new body from sheet metal. This can be done in a small workshop by making a hardwood block model of the inside of a magazine housing. Sheet metal is bent and pounded around this form to make a replacement shell. The real trick, of course, involves getting the top lips of the shell to approximate those on a real factory magazine. Then a catch cut must be filed in the magazine to allow it to be held in the gun.

Magazines disassemble by removing the bottom plate, which allows access to the springs. The sliding piece on the magazine is held in place by folding the steel to create a rim or lip. This is tough to duplicate in home workshops. When making a new magazine from scratch in an emergency situation, I have spot brazed or glued the bottom access assembly to the magazine body. It is probably better to glue magazines together, because it's easier to get them apart again if necessary.

The best instructions I have seen for building magazines from scratch can be found in any of the *Home Workshop Guns for Defense and Resistance* books by Bill Holmes. But even Bill admits that the process is time-consuming, and you'd be better off acquiring surplus magazines for this vital area of proper firearm functioning.

• • • • •

A guerrilla gunsmith will have some severe magazine problems to deal with. Nothing will be quick or easy about this phase of the business, especially when there are no new magazines to look at as guides.

You must check fit and cartridge presentation before lifting a file or tube of glue. Look for worn notches and lips. Take the least intrusive, mildest course of action, working on up through various little options until you have a magazine that will function in your particular weapon. I have no additional advice other than emphasizing that, when it is very important that a gun work properly, you may have to spend great amounts of time making it work properly.

chapter 10
Repairing Broken Stocks

On the lookout for trouble. Note the makeshift but adequate stock repair with duct tape.

REPAIRING BROKEN, SMASHED, AND ABUSED GUN STOCKS ranks right behind unplugging barrels as a principal chore facing guerrilla gunsmiths.

Perhaps it's because guns used for military or paramilitary work receive harsher treatment. Or average users are less well-trained and knowledgeable about guns and therefore treat them carelessly. Or, in the heat of deployment, whether or not the stock shatters is not nearly as important as overcoming an adversary. Obviously I don't really know why there are so many broken stocks to mend, but—take my word for it—there always are.

FROZEN SCREWS

Often when these stocks come in for repair, they will be on ancient guns with rust-frozen screws holding them together. Sometimes the chore of dealing with old stock screws is made doubly difficult by the fact that someone has hopelessly buggered the screw slots. Here is how to proceed.

If possible, disassemble the weapon down to just the screws you must deal with. Soak these screws thoroughly in either WD-40, common automotive brake fluid, or Coca Cola, whichever is the most accessible. Any one of these liquids will do a tremendous job of cutting rust and corrosion, leading to loosening the frozen screws. I know this is frequently out of the question, but if at all possible, wait a few days for the solution to penetrate.

While you are waiting, construct a screwdriver bit. This is a sawed-off piece of screwdriver with a squared shaft that can be turned with a wrench. It is made from a regular screwdriver with a blade that fits or at least kind of fits the screw you are working on. Cut the shank from the handle so it's about 2 1/2 inches long. Securely braze a hex nut onto the small end of the shaft, about 1 inch from the end.

After soaking the screw as long as possible, clamp the screwdriver bit securely in a vise with the shaft end against one of the jaws, the blade engaging the recalcitrant screw, and the gun part

Repairing Broken Stocks

Badly buggered, rusted-in screws on this ancient Marlin rifle required special treatment. After soaking with penetrating solution, we clamped the gun and screwdriver bit in a vise for removal.

on the other side of the vise. Put a thin board between the gun part and the vise jaw to prevent unnecessary marring.

Now tighten the driver bit in the vise, pushing it into the screw securely. Grasp the hex nut with an appropriately sized wrench and turn the screw a quarter turn out. Open the vise a tiny bit and turn another quarter turn. By this time the screw should be loosened sufficiently to withdraw it. This is the only certain way I have ever found to take out rust-frozen and buggered screws, especially when doing stock work.

But here's another method. If the part in question is in a position to do so, precise heat applied to the head of the screw will expand it at a rate different than the rust, breaking the bond. As it cools, introduce a little Liquid Wrench. Then, the fastest way to apply pressure to the screwdriver when the slot is buggered is

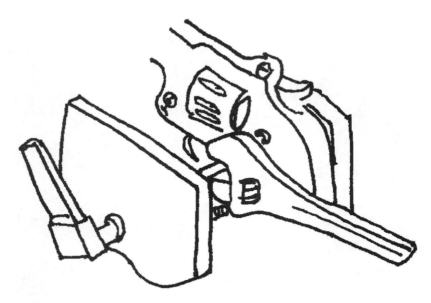

Taking out a frozen screw using a vise and screwdriver bit.

to hold the driver in the chuck of a drill press and the workpiece in a vise. Pull down on the quill and turn the chuck by hand with a wrench. If that fails, drill a hole and use an "easy out" (broken screw extractor).

Also, depending on the screw head and its size and location, you may be able to freshen the slot with a Dremel tool.

GLUING STOCKS

Pieces, chips, and chunks broken off of stocks can be epoxy-glued back together fairly easily so that the repair is not even obvious. Difficulties arise when the stock has been soaked with oil. Epoxy doesn't hold well in oil-soaked conditions; white wood glue doesn't hold at all.

Repairing Broken Stocks

Some 20 years ago, this Marlin rifle stock was split in three pieces. Even today, the epoxy repairs are only faintly visible.

If the stock can be heated gently in an oven set at low temperature or with a propane torch, some surface oil can be dried out of the wood, where it can be mopped up with a rag. Soaking a few days in an alcohol bath is another common method of degreasing a stock. Unless the stock is very old and very oil soaked, either method will work. The main thing is to remove as much oil as possible in the time allotted to this particular project.

An alternative way to leach oil from a stock is to use volatile solvents such as MEK or toluene. Let it soak, wipe and let dry. Let it soak a shorter time, wipe and let dry. Let soak a still shorter time and wipe again and let dry. After that, the surface and substrate should be clean enough to accept an epoxy bond. Do not leach with acetone, as it will dissolve the cellulose of the wood.

Guerrilla Gunsmithing

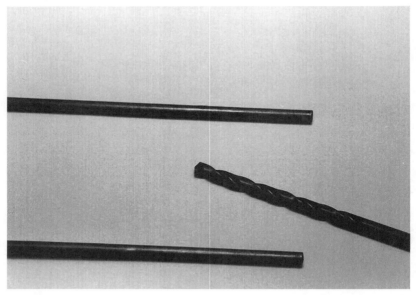

Brazing rod pins (1/16 inch) and drill used to set pins to repair badly broken stocks.

If you have all the broken bits or pieces, simply coat them with epoxy and clamp them back together. Wait two days for the adhesive to cure. Then, if appropriate, the stock can be refinished. Remember that rifle stock that was destroyed by rampaging hogs 20 years ago? The repairs I made on it are still functional today and not at all obvious.

PINNING STOCKS

Pinning stocks is a frequent necessity in a guerrilla situation, especially when damage is extensive or is in a location on the stock that is especially sensitive, such as the wrist. But rather than resorting to wooden dowels, use small lengths of 1/8 inch brazing rod. For some strange reason we forget

Repairing Broken Stocks

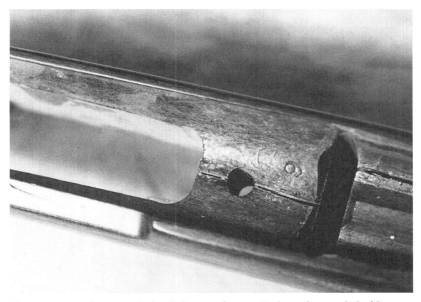

Even some very large cracks in vital areas of a gun stock can be mended with epoxy and brass pins.

about brazing rod even though it is far stronger and easier to deploy than common wooden dowels. Not only is brazing rod far stronger, but the insertion holes required are smaller and less destructive to the stock, allowing braze pins to be inserted in places where there is no room for a wooden dowel. Common wood screws could also be used, but I've found that they don't always reach in far enough and tend to further split the wood.

Pinning works nicely on both wooden and solid composition stocks. (Composition shell stocks as found on AR-15s must be repaired by other means.) It is especially valuable when dealing with a green stock break or a small twist-and-turn fracture as seen on some guns that have been run over by a vehicle. In this case, the stock is twisted but not broken completely away.

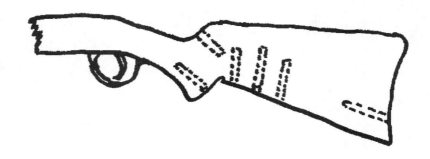

Suggested placement of brazing rod pins in damaged gun stock.

Start the repair process by straightening out the stock back to its original configuration as much as possible. (Completely broken stocks are jammed back together and held temporarily with epoxy.) Now drill two or three small holes lengthwise into the break, just of sufficient diameter to accept whatever size pins you have on hand. Space these holes about 1 inch apart. Narrower spacing isn't as effective because the wood loses too much of its inherent strength.

Run the pins in to be sure they fit and are of the correct length. Now epoxy the whole thing together. Be sure to get lots of glue worked into the break as well as into the pinholes around the braze rod pins themselves. In this way, complete fractures can be mended even with a single pin and lots of epoxy.

Pins can be inserted just about anyplace in a stock that doesn't impinge on the gun's working parts. A single brazing rod pin up the pistol grip, for example, is amazingly strong. Keep in mind this is not finish gunsmithing. We want the gun to be functional, not beautiful.

FIBERGLASS REPAIRS

Badly fractured stocks and forearms can be repaired by wrapping them in fiberglass if this seems easier and more practical under your circumstances than pinning or gluing. External wraps of fiberglass can be very strong and actually is the preferred remedy when stocks are crushed to the point where they pretty much must be rebuilt.

Shards from dried fiberglass can cut hands, fingers, and faces, so after fiberglassing a stock it is helpful to sand rough edges and then paint over the whole thing with something or other. In one notable case, we used old nail polish to cover a glass-repaired stock; in another we coated it with floor varnish to keep shards in check.

When making external fiberglass repairs, consider positioning a brazing rod pin or two on the outside of the stock as a kind of splint. These pins will add tremendous strength and can be bent and molded somewhat to follow the contours of a gun stock. External pins are especially valuable when repairing composition shell stocks such as those on AR-15s, as normally these cannot be repaired from the inside to any great extent.

WRAPPING

Shell or solid stock, good, effective repairs can be made with something as quick and easy as long leather or nylon shoelaces. Again, place brazing pins along the break on either side to act as a splint. Starting far back behind the break, wind the laces around the stock. Work evenly and pull as tight as humanly possible. Completely cover the external pins with the layer of lacing. After wrapping the stock and pins, coat the whole thing with paint, varnish, or whatever glue you might have. This coating really locks down the lacing and prevents it from stretching and loosening. Of course, the entire gun will be bulked up in the process, but it will still be quite usable.

Guerrilla Gunsmithing

Rifle and shotgun stocks can be repaired with lace or even rawhide when it is the only thing available.

Wire, string, or lace wrapping is usually a last-ditch emergency measure undertaken when the gun must be put into action quickly and when no other materials are on hand. Wrappings do not have to be coated to work, but they will last longer if stabilized in this manner.

One time in Africa years ago, under absolute emergency circumstances when we had no other materials, we wrapped a broken Egyptian Hakim rifle stock with duct tape. It has been a long, long time, but as long as everything else was properly cared for, I see no reason why the rifle wouldn't still be functional today. If nothing else, it may have bought the owner enough time to scrounge up a replacement stock for his gun.

In yet another emergency case that vividly illustrates the extent to which guerrilla gunsmithing must go, we snared a small female

Thompson's gazelle, from which we took a piece of rawhide. We laced this green hide around a break in the wrist of an SKS rifle stock. After it cured, dried, and hardened, the rawhide-strengthened stock seemed as strong or stronger than the original. Keep this concept in mind when there is no epoxy, fiberglass, or duct tape to be had. Rawhide makes a wonderful, smooth, strong repair. It virtually glues itself to the stock!

MAKING STOCKS

Stocks for some guns are not particularly difficult to make by hand in regular home workshops. Two that immediately spring to mind are those for the AK-47 and the Russian RPD light machine gun. Both stocks can be turned out by hand using little more than a handsaw and wood rasp. One time in Somalia we made an AK stock out of a piece of nondescript native 2 x 6 lumber using only a bench grinder and machete!

Making the stock wasn't that difficult, even though we had so little to work with. Those who have the time, along with a few basic hand tools such as drawknife and wood rasp, should consider this option, especially if it's one of those simple Soviet Bloc weapons encountered in such large numbers around the world. Consider using a combination of new wood along with fiberglass or epoxy to cover errors and to hurry the project along. It helps immeasurably to have at least some of the old stock on hand to use as a pattern.

Again, Bill Holmes' *Home Workshop Guns for Defense and Resistance* series contain much valuable information on making wooden stocks from scratch, including diagrams and dimensions. Although his instructions generally require machine shop tools such as a drill press, the resourceful guerrilla gunsmith will be able to adapt.

FIXING SHELL STOCKS

Most AR-15 type nylon shell stocks are reparable, even those that have been badly crushed, blasted, or broken. As mentioned,

Guerrilla Gunsmithing

these can be stabilized with pieces of external brazing rod and fiberglass. The problem is they really look crummy after they are fiberglassed or glued from the outside. I don't mind an old Enfield with extensive wrappings, glassing, and repair, but an AR-15 with a big patch looks obscene to me. On the plus side, these types of stocks hold glass and epoxy much better than wood.

It isn't difficult to take shell stocks down to work on them as best you can from the inside, although results may be marginal at best. Again, use fiberglass or epoxy. Be sure none of the glue or glass gets into the recoil assembly or other working areas of the gun. After shoring these stocks up on the inside, it will probably be necessary to slather on the glue or glass on the outside, depending on the extent of the damage.

• • • • •

I, or people I have worked with, have repaired absolutely every busted stock we've encountered. Some were a real sorry mess, having been run over by an APC or truck. Others had been intentionally broken by enemy soldiers in an attempt to render the weapon useless.

In the case of one AK-47, the stock was busted and the rear of the receiver was bent and twisted as a result of government soldiers trying to reduce the gun into unusable scrap. We beat the soft steel receiver back into shape. Then, using the old stock pieces as a pattern, we roughed out a replacement in part using a carpenter's skill saw. In another notable case we repaired the inside of an M1 carbine stock with epoxy and the outside with dozens of rubber bands! The gun worked just fine in spite of our makeshift improvisations.

The important concept here is to realize that you, as a guerrilla gunsmith, will encounter a great many broken stocks. It is best in these circumstances to think outside the box. Use whatever materials and means are at hand to get the gun functioning as quickly as possible.

chapter 11

Making Small Parts

A great many common pieces and parts can be formed into gun parts.

I have already mentioned the fact that it may be necessary to make some small replacement parts for guns in a guerrilla gunsmith situation. By definition, these parts must be made with a minimum of tools and supplies.

As a general rule, guns made before the turn of the 20th century contain more extensive milled parts that are difficult to replicate. Pity the poor guerrilla gunsmith who is called on to make replacement parts for, say, a really old Mauser pistol. I am reasonably certain it is not possible. Later in the 20th century the trend moved toward guns made with simple stampings or stock, off-the-shelf parts, which are easier to replicate or substitute.

(When I was doing this sort of work some 20 years ago, the guns with which I had to deal with inevitably were old Enfields, Springfields, Nambus, Arisakas, and Nagants. These were the weapons most likely to turn up in the hands of rural, remote natives, and they seemed to require the most repair and maintenance. Perhaps because AKs and SKSs hadn't been around as long back then, or perhaps because parts and repairs are easier for those more modern weapons, I was seldom asked to work on them.)

FIRST STEPS

The first rule of thumb when confronted with the need for small replacement parts is to see what you already have on the shelf that could be utilized immediately or modified into a part for a gun. Obvious small parts include springs, screws, and bolts, though many will not work without extensive remodeling. Remodeling may entail shortening, reworking the heads of bolts, cutting notches, or whatever else that's necessary to make the part work on a gun. Larger, off-the-shelf parts may include hunks of muffler pipe, tractor parts, sections of cold-rolled steel, and other miscellaneous bits and pieces you have accumulated.

In all cases, modifying existing nuts, bolts, screws, springs, tubes, and steel is much easier than starting from scratch to make new ones. It is also a fact that guns are increasingly being assem-

Making Small Parts

Power tools make the job of producing gun parts much easier, but most of the time the guerrilla gunsmith will have to do the work by hand.

bled with lots of stock, off-the-shelf parts that you may be able to scrounge up before attempting to make your own.

The second rule of thumb recognizes that some people, especially those in the Third World who are accustomed to improvisation, are incredibly skilled at knowing how a missing part should look and act within a gun's mechanism without ever actually having seen it. Some of these people can study a gun's innards and know immediately how to recreate a missing or broken part. I am certain, for example, that in several instances when we were rebuilding triggers on FN rifles, the skilled artisans with whom I worked had never seen the gun or the parts we were attempting to recreate. Yet they were able to visualize and then draw these parts on paper in a most remarkable and accurate fashion.

The moral of the story is that if it is at all possible, search out these people to assist with your specific guerrilla gunsmithing project. It is helpful to merely know that most small parts can be made by hand with primitive tools, but it is better if you can enlist the help of one of these skilled artisans to actually do the work. In many cases their abilities are almost magical.

FIRING PINS

Of the small parts you may have to make for yourself, the most common are broken firing pins. Fortunately, firing pins are relatively easy to duplicate. Usually most or all of the broken pin is retained by the gun. As a result, it is easier to determine critical lengths and diameters. As a practical matter, make the pin a bit long, then shorten it with a stone or file as required to actually fire the round.

Start the process of making a firing pin by selecting a hardened bolt as close to the size and length of the pin as possible. Hardened bolts are those with the marks on the head indicating levels of strength (for gunsmithing work, use at least a grade 8, indicated by six radial lines on the head). They make excellent firing pins.

Anneal the bolt to take out its hardness so that it can be

Making Small Parts

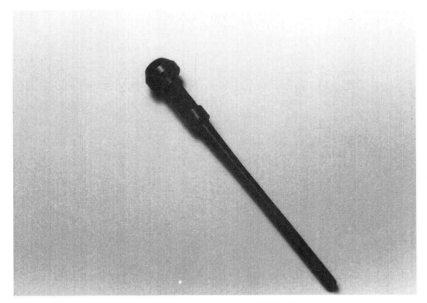

Firing pins are the most common small parts that fail in firearms.

more easily filed and turned down to size. Chuck the bolt into your electric drill or drill press and go to work with a file. Keep shaping it down until it is the same length and diameter as the pin you are replacing.

Firing pins are relatively straightforward. All it takes is lots of patience, especially if you had to start with an overlarge bolt or you have not annealed it properly. Having a micrometer to measure your work is very helpful. If not, simply take the pin out of the chuck and try it in the gun's bolt from time to time.

Expect to spend three to four hours "tooling" on the old bolt, especially if you have nothing but an old file with which to do the work. Be especially careful when working on relatively long, thin firing pins that you don't bear down excessively so that the pins starts to wobble off center.

Bolts make good material for gun parts. The hash marks indicate degree of strength. Three radial lines indicate a grade 5.

MAKING PARTS FROM SCRAP STEEL

Small parts that often break or become lost, such as ejectors or extractors, can be fashioned from heavy stainless wire. It will take a bit of experimentation and fooling around, but they can also be reformed from stock spring steel wire or from very thin drill rod. The biggest challenge may entail finding sources of supply for the wire. As previously mentioned, try aircraft and auto mechanics or virtually any manufacturing or processing facility in the region.

Other essential parts that have broken or become lost, including pieces like magazine catches and grip safeties, are a real challenge. Start by drawing an outline on heavy cardboard of how you think the part looked. Cut it out and slip it into its

Making Small Parts

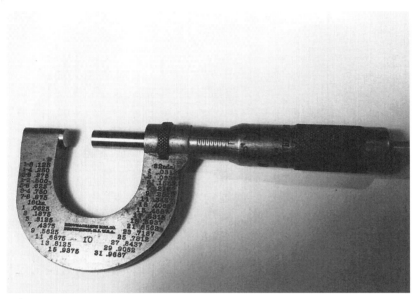

Guerrilla gunsmiths should learn to operate an old-fashioned micrometer.

allotted place to see if the function seems appropriate. Keep cutting and snipping on the template until you have something that looks usable. Then lay it out on the scrap material you will use to replicate the new part (try to pick steel that you think approximates that in the original part) and draw an outline of the piece. This scrap can range from a worn wrench to the frame of a wrecked auto.

Rough out the piece with your file, hacksaw, bench grinder, Dremel tool, or whatever. Of course if you still have pieces of the old, broken, or worn part, it is much easier to envision how the new one might be formed from the scrap on which you are working. Keep in mind that some parts can be produced flat and then bent to fit after they are roughed out. Other times it is simpler to make parts as two components that are then soldered,

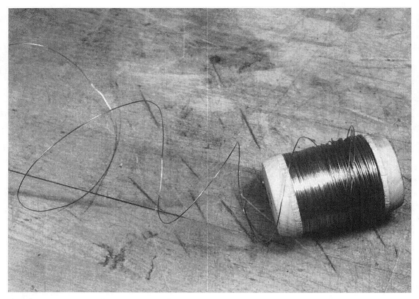

Heavier grades of stainless wire make good material for springs.

brazed, or glued together. It isn't always possible, but at times it's the only way to make a tough, three-dimensional part.

Common drill presses can be turned into quasi-milling machines by the addition of small, adjustable tables and use of a milling head in the chuck. This is a very important option to keep in mind when contemplating roughing out complex parts. Files will get the work done eventually, but using a drill press and mill may get it done a day sooner.

When using a mill in a common drill press, be very careful that the steel is annealed and that you work the mill in and around the part slowly and cautiously. Take off only very small amounts on each pass. It's a freehand procedure, but I have seen some skilled machinists duplicate some very complex parts completely by eye with this method. In one or two instances,

Making Small Parts

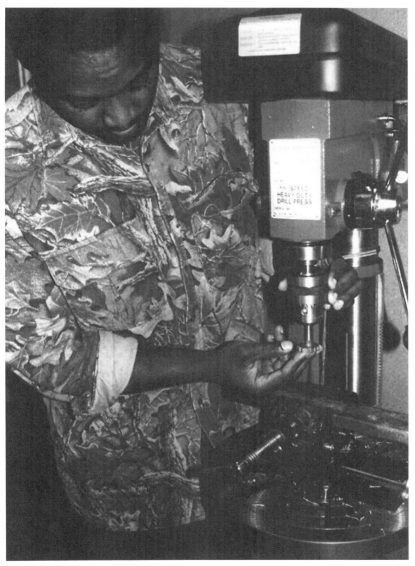

Inserting a common mill into a drill press in preparation for milling a piece of mild steel in the adjustable table.

Guerrilla Gunsmithing

Setting up an adjustable drill press table to perform simple milling operations.

Making Small Parts

craftsmen actually made the parts somewhat differently so they would hold up better under rough conditions. At least that's what they told me, but the results in service seemed to vindicate their "field engineering."

It is common practice to use a hacksaw to rough out flat parts, which are then filed down to exact size. When starting with a hacksaw, don't try to cut out the part right on the line you've traced. Leave a little metal as a border that you can then file and adjust precisely.

• • • • •

Making replacement parts from scratch entirely out of scrap metal is possible. The problem is that this brief chapter makes it sound easy. It's not. Most small parts require a day or more of finger-fatiguing work that is very, very difficult in its execution. But the important lesson is that we know it *can* be done and we keep at the project, never giving up until satisfactory results are achieved.

Pakistani artisans in the Khyber Pass area, for instance, turn out modern weapons of most models and designs using little more than hand tools. They deploy some electric tools such as drill presses, bench grinders, and Dremel-like devices, but none of these are out of reach of the resourceful guerrilla gunsmith. Some complex work such as rifling barrels is contracted out to more sophisticated shops among the group.

Although this should be an inspiring example for all of us, remember that these people work at the business 300 days a year. As a result, they become very skilled and relatively speedy at what they do, attaining a skill level greater than we as part-time guerrilla gunsmiths can probably ever hope to duplicate. But we *can* learn from and be motivated by their example.

Conclusion

At the time the U.S. Constitution was being written and put into place in the 1780s, there was great debate regarding its exact wording, long-term implications, and underlying philosophies. Thousands of pages of editorial, testimony, and opinion regarding our various rights, including the right to keep and bear arms, were produced, and the opinions of famous and not so famous men are still with us today. Much of this material is worth reading again.

From the January 7, 1788, editorial section of the *Courant*, a Hartford, Connecticut, newspaper:

> *"In countries under arbitrary government, the people oppressed and dispirited neither possess arms nor know how to use them. Tyrants never feel secure until they have disarmed the people. They can rely upon nothing but standing armies of mercenary troops for the support of their power. But the people of this country have arms in their hands; they are not destitute of military knowledge; every citizen is required by law to be a soldier; we are marshaled into companies, regiments and brigades for the defence of our country. This is a circumstance of the people; and enables them to defend their rights and privileges against every invader."*

And again, in the *Freeman's Journal*, Philadelphia, March 5, 1788:

> *"The freemen of America will remember that it is very easy to change a free government into an arbitrary, despotic or military one: but it is very difficult, almost*

impossible to reverse the matter—very difficult to regain freedom once lost."

Obviously, I am not optimistic regarding the future of freedom in the United States. As of this writing, major politicians openly campaign on the promise of doing away with our right to keep and bear arms. These are not stupid individuals; they must know that their actions will lead to despotism, and apparently this is the direction in which they wish to proceed. On the other hand, a certain percentage of the citizenry seems intent on playing into their hands by trading freedom for some perceived security.

Preserving our freedoms won't be easy. It is going to be a difficult, do-it-yourself project. It is this philosophy that drives the guerrilla gunsmith, who realizes that, no matter how difficult the chore, he must retain some sort of operational weapon in his hands to defend against the inevitable tyranny of an omnipotent State.

Like all of survival (or "preparedness" for those more comfortable with that term), we are going to have to handle our gunsmithing chores ourselves when there is no other alternative. In most cases we would be better served by paying a full-time expert to repair and maintain our arms. Guerrilla gunsmithing can be a very inefficient, time-consuming undertaking, as is most of survival.

I believe we have covered the most basic of problems related to keeping our guns operational under tough circumstances. Those who have additional thoughts and suggestions can contact me in care of Paladin Press. There are bound to be subjects that I did not think to touch upon or that did not come to my attention during several decades of guerrilla gunsmithing. It is becoming a matter of life or death that we have this knowledge—as well as the underlying philosophy that drives it—in as complete a fashion as possible. In this regard alone, I trust that I have struck my own little blow for freedom and that you will find this information helpful.